Managing People in the Hvac/r Industry

Howard J. McKew, P.E.

BNP
Business News Publishing Company
Troy, Michigan

4/96

30976516

Library of Congress Cataloging in Publication Data

McKew, Howard J.
    Managing people in the HVAC/R industry / Howard J.
McKew.
        p. cm.
    ISBN 0-912524-97-9
    1. Heating and ventilating industry--Personnel manage-
ment. 2. Air conditioning industry--Personnel management.   I. Title.
    HD9683.A2M35   1995   94-33085
    697'.0068'3--dc20   CIP

Editors: Joanna Turpin, Carolyn Thompson
Art Director: Mark Leibold

Printed in the United States of America
7 6 5 4 3 2 1

Acknowledgments

Management is a medium through which you can take control of your job, career, and personal fulfillment. I first began as a drafting trainee and have worked with some very good leaders throughout the years. I have absorbed the teachings of these leaders to formulate my own abilities and style of management. The leaders I mention in this book are the select few who were very good at what they did.

Not mentioned in this book are my wife Joyce and my children Amanda, Dan, and Kim. They are the reasons why I enjoy what I do. They are the reward at the end of my day.

This book is dedicated to my parents, Charles and Carmelita McKew.

About the Author

Howard J. McKew has been an engineer for over 28 years and a manager in the hvac/r industry for over 15 years. He is a registered professional engineer, as well as a member of the following organizations: National Society of Professional Engineers; American Society of Heating, Refrigeration and Air Conditioning Engineers; National Fire Protection Association; and American Society of Plumbing Engineers.

Mr. McKew writes a monthly column for *Engineered Systems* magazine and is currently vice president-engineering with William A. Berry & Son, Inc.

Table of Contents

Introduction

Successful heating, ventilating, air conditioning, and refrigeration (hvac/r) management can be compared to directing traffic on a two-way street. The hvac/r manager must keep the flow of traffic moving smoothly in both directions by directing from the middle of the road. The manager must look out for both the company **and** the employees. While the two can go forward from different directions, the manager must be the "traffic cop" responsible for their advancement and success. Employees often believe they are on one side of an issue and that management is on the other. Being fair to both sides and making decisions in the best interest of the corporate mission statement is the responsibility of the successful manager. This book is a practical, on-the-job guide to managing people in the hvac/r industry. Over the past 17 years, I have compiled a number of guidelines, phrases, standards, and philosophies of managing people that are based on what I have learned and what I know has worked for me.

Managers can maximize their success by recognizing the value of **time management**. Implementing time management will give a new manager the tools to handle effectively all of the responsibilities involved in the managing of others. Those who are put in charge of others are in a middle management position. This middle management position is the role of an effective manager. Whether you are truly in the middle of a corporate organizational chart or at some other level, this position description can apply to anyone who has responsibility over others, including foremen, senior project engineers, vice presidents, or even presidents. These people should consider themselves spokespersons for the company and its employees and perform in an equitable manner to all. The phrase I use to describe this role is "looking out for both sides." If you are to be a successful manager in a successful company, you must conduct yourself in this way.

I officially began my management career in 1977, when I was given the opportunity to assist the department head of a mechanical and electrical consulting engineering firm. There were approximately 34 people in the hvac department, and I grabbed the chance to become the assistant department head. At the time, there were numerous issues within the company I believed needed changing. The good news was the consulting firm was experiencing growing pains. At the same time, a high employee turnover rate was affecting the work, and the atmosphere was one of disenchantment. When the company was smaller, it was able to attract draftspeople, designers, and engineers, and there was also a sense of closeness. Success created change, and those changes weren't always for the good of the company or the employees.

The frequency of worker turnover was a concern of mine, along with the need to improve employee morale. The frequent loss of in-house engineers, designers, and draftspeople, in addition to the inconvenience of bringing in new replacement personnel, made it difficult to maintain the company's high standards for producing excellent construction documents. Production engineering efficiency had dropped noticeably. My goal was to help improve the quality of work and the working environment, which, in turn, would have a positive effect on engineering performance.

Having had some success assisting in the management of the hvac department and its personnel, I wanted to take on a more responsible position as the manager of engineering. It seemed to me that management needed to be doing a better job, so I pursued, and received, a promotion to manager of hvac engineering and was responsible for the work of approximately 40 employees. Today, I can't seem to get out of management! The role has stayed with me, while the responsibilities have expanded from design engineering to include operating and maintenance engineering, environmental control engineering, estimating, and construction management.

Having once been described as an introvert trying to be an extrovert, I sometimes wish I could go back to a more singular job description. However, management does offer the opportunity to make changes, achieve results, observe employee job satisfaction, and be part of a company's success. These things, along with learning what works and what doesn't, make the job worth the time and commitment. These and other related experiences have provided me with a wealth of useful short stories, which illustrate what worked for me and what didn't. As a person who has walked the "middle of the road" in business, I have often shared my experience with others and used it to become a better manager. These matters and other similar topics are quite common in most management operations, not just consulting engineering firms. By putting into words what I have learned on-the-job, I hope to help others as they venture into and through the management ranks of both small and large companies.

Writing comes a lot easier to me now, as compared to my first attempt. In 1975, I was encouraged to write an engineering article for publication. At the time, I was working for an engineering and architectural consulting firm, which paid employees $250 for writing a technical article and having it published. I don't know what it was that made me think I could write, but I had this title "Energy Conservation Through Value Engineering" in my head and the company encouraged employees to take a chance. Although it took me 18 attempts to get it right, I finally succeeded in putting my experience into words with the publishing of this article in *Heating, Piping and Air Conditioning* magazine in September, 1976. Even more surprising, the article was reprinted in a book of energy design concepts in 1977.

To the best of my knowledge, I was the only one in the company who took advantage of the writing reimbursement policy. This was unfortunate, because I learned a number of valuable lessons from my first writing experience. First, I learned more about what I had engineered by putting it into writing. People often fail to think a task through when working on a project. When you have to put it down in writing, the task becomes clearer to you and to others. This is an integral step in the process of continuously improving quality, and it is well worth learning. In addition, you are an expert on the topic you are writing about, because it is your project. You are the authority. You know the most about that task, job, responsibility, etc. Through years in management, I have found that writing about my on-the-job experiences comes easy, because I am the authority on my experiences.

A second lesson I learned was to take good notes! Long before I took my first shot at writing, I had the opportunity to work with a person I consider to be the best production manager in the business, Mr. James McGrath. Jim was a stickler for detail. His job was to manage the company's design engineering activities. As a trainee in the firm, I was fortunate enough to have him oversee my work. In the process, he taught me to take good notes, keep a "things to do" list, and maintain a sense of humor. My note taking became proficient and, as a result, made my future writing efforts easier. Taking notes has now become a habit with me. Likewise, "things to do" lists have become good time management tools, while maintaining a sense of humor is a must in work and in life.

The third lesson I learned from my first published article was that I possessed creativity. Since grammar school, I have had the ability to draw reasonably well, and that talent offered me the opportunity to express my creativity. Beginning with my first published article and with each subsequent article, the title provides me with the inspiration to write. A catchy title gets me started; good notes provide me with the information to complete the story.

Being able to express this creativity in a technical business environment was a unique opportunity. Mr. Edward Shooshanian, President of Shooshanian Engineering Inc., has always been an inspiration to me. He allowed each employee the freedom to do the best he or she could do as a design engineer. I was able to grow intellectually with a supportive boss, rather than stagnate within a bureaucratic environment. In later years, I was exposed to companies with stringent policies and procedures, which inherently limited the creativity of their managers and employees. In order to keep both styles of management in perspective, I often remind myself of a phrase from the book *Robert Kennedy: Promises To Keep*, which says, "Some men see things as they are and say, why? I dream things that never were and say, why not?"[1] This questioning of "why" and "why not" is essential to having creative ideas, creative environments, successful companies, and successful individuals.

Being a creative leader/manager takes years of experience, commitment, and a lot of extra work. I do not believe anyone can get ahead working 40 hours a week, because you need to do not only your job but learn the next job as well. Demonstrating that you can get the job done within the scheduled time frame is a must. Learning the skills needed to move up the ladder requires time management, creativity, determination, and a "can do" attitude. In the process, a successful administrator will become a spiritual leader to the employees and to the company. It is that person's job to get the most out of each employee, for their benefit as well as that of the company. The "real world" management experiences I relate in this book have been part of my on-going process to continually strive to reach this goal.

Having observed how management functions, as both an employee and a manager, I have tried to note what works and what doesn't work. I have always done what I thought was right (not necessarily what was popular) and have tried to learn from others. In addition, I have developed numerous management "tricks of the trade" and personality profiles. With more than 17 years of management under my belt, I have developed my own rules by which to manage. I have also observed similarities in people and categorized their activities into specific types of individuals/workers. This has been very helpful in the management process. I believe a successful manager needs to be cognizant of the various types of personalities when looking out for the company and looking out for the employees. Sometimes the correct thing to do is in favor of the company, and sometimes it is the employee who is right. Knowing how to present the decision to the individual and planning your directive based on how you anticipate their reaction, is a strategic part of managing people.

The two-way street management philosophy places everything on a level playing field. This philosophy is invaluable when considering a company's performance and success, ability to add and reduce staff, and employee performance and reviews. This method of management is also helpful when assessing the success of individuals, providing direction to these individuals, helping individuals set goals, and managing the process fairly. Performance reviews are probably the biggest problem for managers. Reviews are one of the most important issues to employees, yet managers seldom appreciate their significance. Planning and addressing an employee's performance is a daily, weekly, and monthly process that is formulated into an employee's annual review. After approximately 500 one-on-one meetings with individuals, I believe this is the foundation of good employer-employee relations.

This book looks at the sense of accomplishment people can experience when they have been the best managers they could possible be. Successful managers must recognize their value to the company, their family, and themselves and must learn to keep "all of the balls in the air." However, juggling these three important issues can be both sweet and bitter sweet for many managers.

I have structured this text so each of the chapters can stand alone as a separate entity. After you have read the book, you can quickly refer back to any one of the chapters and easily locate the information you need. I have also included some standard forms that you should find useful when managing people in the hvac/r industry.

NOTES

1   *Robert F. Kennedy: Promises to Keep, Memorable Writings and Statements,* Hallmark Edition, Kansas City, Missouri, 1969.

Chapter 1
Time Management

Time management is the cornerstone to management success. Most managers have a large number of people to manage and/or numerous job responsibilities. At one point in my career, I had 40 engineers, hvac/r designers, and draftspeople to manage. At the same time, I continued to be responsible for the engineering of in-house projects. All through my management career, I have stayed involved with the design of hvac/r systems. As a person who has moved up through the professional ranks, I believe it has been important for me to stay active in the arena where I have excelled. Doing so has required maximizing the time I spend at work. I am sure this is true for many of the people in management. Staying abreast of the changing technology within the hvac/r industry is vital to success. Good time management skills will allow you to achieve this goal.

Time management skills are also an integral part of being a good draftsperson, designer, service technician, etc. A draftsperson must be proficient with the information he or she documents on the drawings. Carefully noting the work that has been completed and identifying what remains to be done demonstrates the need for these good organizational skills. Service technicians need similar skills to be proficient at their jobs. Minimizing the trips to and from the service van and supply house can mean the difference between a good job and a very profitable job. Employees must maximize their time if they are to complete a project on schedule and at a profit. This also helps employees with their goal to continually prove themselves worthy for advancement considerations. Through successful management of time, a person can consistently complete an assignment within the prescribed time and estimated cost.

At one particular firm, I was assigned the responsibility of managing all of the engineering and estimating duties. I was given these additional responsibilities because of my time management skills. Managers responsible for multiple groups must schedule sufficient intervals of time in the workday to keep all the "balls in the air," and time management will prove to be an invaluable tool.

YOUR TIME IS IMPORTANT

Time management means being in control of yourself, your work, and your life. I often use the example of a person juggling three balls at one time. Although you can put these balls in any order, because they are all equally important, the first ball is your work. In order to be successful at your trade, you must function at peak performance consistently. Managers and employees must always be cognizant of their effectiveness and ability to get the job done. At the same time, they also owe it to their families to share their life and responsibilities at home. Keeping this second ball in the air is just as important as the first. Equally important is having time for yourself. Whether it's a six mile jog, a good book, or painting a picture, people need to get away from the demands of life. Keeping all the balls in the air is essential to a manager's success. You must be efficient at work and at home to be truly successful in life, and time management will help you to be efficient.

Managers must also be in control of their time, if they are to be responsible for the time of others. They must be able to juggle their busy work schedules with their personal lives. I don't believe you can manage others effectively if you can't control your own daily routine. This may sound old-fashioned, but the top management performers I have known were all successful at keeping all the balls in the air. Making the time to meet your commitments, both personal and professional, is equally important to personal satisfaction and confidence. As a manager, being aware of how employees are balancing their work and personal lives can be an important way to maintain work that is on schedule and without defects. If an employee's job satisfaction conflicts with his or her personal satisfaction, the product may suffer. Time management can smooth the waters and guide people back on track with their responsibilities.

Maximizing your time begins early in the morning, because that's usually when the workday starts for those in the hvac/r industry. If you are not a "morning person," then chances are this business may pass you by. I realized early in my career that I performed best early in the workday. I'm not saying that work didn't get done late in the day, but my most creative work seemed to come to me before the routine chores of the day took over my schedule. Each person should stop and concentrate on when they believe they are at peak performance. Often workers will be late starting and late finishing. There is nothing wrong with this concept if it is in concert with the company's operation.

I recall one manager who seldom arrived before 9:00 a.m. He would work until 7:00 or 8:00 p.m. and was at peak performance by early afternoon. This can work in some firms, but it didn't work here. The group this manager was responsible for started work at 7:30 a.m. His schedule also didn't meet the customer's needs either, because he worked too late in the day. As the manager, it is difficult to be in charge if you are the last person to arrive in the morning, especially since you are considered a leader and role model for the other employees. The workers interpreted this manager's behavior to mean that they could slack off until he showed up. At the same time, their workday ended at 4:30 p.m. Many employees do not appreciate or even care whether you work additional hours. They can attribute your staying late to any number of reasons that don't affect them, such as not getting the job done on time or your high salary.

Time management specialists must recognize which time of day is their peak performance period. It is important that you minimize office interruptions and optimize your work output during this time period. Disruptions can come in many forms, i.e., phone calls, people asking questions, requests to leave the office to attend a jobsite, meetings, interviews, etc. Managers can remain in control of their work schedules by being aware of distractions and being prepared to handle these requests for their time. With each request to meet outside the office, a manager can effectively reschedule most meetings without upsetting a customer. I have found that some customers are aware of this deficiency and will accommodate a request to have the meeting at the start of the day or near the end of the day. On other occasions, the client may not have given the midday meeting time any thought, and he may appreciate the suggestion to make better use of both your time and his by not breaking up the workday with a meeting outside the office.

Arranging out-of-office meetings first thing in the morning or last thing in the afternoon will be a better use of your time. A time management rule I live by is to never leave the office in the middle of the day to attend a meeting. After spending time traveling to your workplace, the last thing you want to do is leave to attend a meeting outside the office. The time spent going to this meeting and returning to the office will usually consume one to two hours of your work schedule. In an eight-hour workday, this trip could consume approximately 25% of your day. Certainly, this is time that could have been better spent.

When in your office, also make a point of not stopping what you are doing each and every time someone walks in on you. This is particularly important after you have recognized the period of the day when you are at peak performance. I believe a manager **should** be accessible, as opposed to those managers who prefer to work behind closed doors. However, an accessible manager must still be in control of his or her workday. In my experience, I often have workers come to my office and interrupt me with a question, request to meet, etc. I don't look up at the person immediately but continue with what I'm doing. At an appropriate moment, I will ask the person if I can help them. Once I know what the person's need or request is, I will usually try to direct the solution back to the person in the doorway or arrange a time to discuss the issue. Seldom will I stop what I'm doing to solve the problem. In time, people will realize that I'm apt to answer their questions with a question, which will result in the problem staying with them. They will also learn that it is often better to arrange a meeting beforehand.

Your time is important! Recognize the peak performance periods of both your workers and yourself. Tutor others in managing their time, both at work and outside of work, for maximum performance results. Don't be afraid to share this strategy with your customers. Everyone appreciates a successful person, and everyone wants to work with a winner. Time management will be the cornerstone to achieving that goal.

TRAINING

When you are out of the office, be prepared to waste time! That's what can happen unless you're prepared to maximize this opportunity for private time. Each day when traveling to work, a person can take advantage of this private time by listening to audiocassette tapes. These tapes are also valuable if you travel by train, bus, or subway. I usually listen to educational tapes at least three times a week while on my way to work. At the end of the day, I prefer a break from my business schedule and usually listen to the news or music. Managers should recommend the establishment of a corporate library that includes audiocassette tapes as well as instructional videotapes.

When developing this library, create a "master list," which inventories the tapes available for employees to borrow. Post this listing for all the employees on a monthly basis. You may even consider sharing this listing with your customers for their benefit and information. The investment is minimal when compared to the benefits received from listening to these educational tapes. In addition, everyone who is interested in increasing their knowledge can have access to and share in this library.

There are a wide variety of hvac/r-related tapes for an employee to listen to while in transit. When purchasing tapes for a library, the topic of time management should be at the top of the list, along with developing a positive attitude, hvac/r service, designing and building, selling, etc. Tapes relating to supplemental concerns, such as quality, dress code, and safety are also valuable and are worthy of use in the library. Use the technology that is available to take advantage of these learning tools. There are hundreds of informative tapes that will help managers and employees with their jobs. Other tapes will complement your learning with information associated with, but not directly related to, your trade.

Video training is another means of **continuing education**. The technology used in the hvac/r industry continually changes, and people in the business cannot afford to miss the opportunities these changes offer. In fact, the phrase "a picture is worth a thousand words" takes on new meaning with video. Professionally prepared videos can be watched during an employee's lunch break, or they can be taken home and watched at the employee's convenience. Managers can also generate their own training tapes and videos.

In the '90s, corporate training will become more commonplace in hvac/r firms. With the quest to improve quality, companies will strive to standardize their processes wherever possible. In-house training tapes, both audio and video, offer a company the opportunity to produce a **corporate lesson plan** that can be used over and over. Instead of "reinventing the wheel" each time a manager wants to teach a specific course, these training tools will already be in the library. Tapes detailing the company's policies and procedures also can be in the library, ready for use when a new employee arrives. Prior to the first day of employment, a video can be given to new employees so that they become familiar with the firm's history, mission statement, achievements, and strategic and tactical goals. These are just a few of the valuable points of interest that can be included in the corporate library.

Today's technology offers many different time management tools for training purposes. Audiocassettes and videotapes are two of these tools for educating the hvac/r workforce. Firms and managers who do not take advantage of these simple and cost-effective devices miss the opportunity to educate themselves and their employees. A manager needs to be creative with the training of his or her staff, the time required to train the staff, and when this training should occur. A company cannot afford to consume the normal workday to educate its employees. At the same time, a company cannot always pay to send individuals to educational classes and still keep overhead costs low. By using audiocassettes and videotapes, a manager can address many training needs through a standardized process, as long as individuals have access to a corporate library. The initial investment easily can be less than a thousand dollars and can grow progressively through an annual education budget.

Many of the initial audiocassettes and videotapes address the basics relating to the work performed by the individuals within the company. If the firm provides consulting services, the tapes may cover more advanced topics relating to engineering fundamentals, such as psychrometrics. If the firm is a maintenance company, then many of the "how to" films available would be appropriate, such as how to lubricate a pump. Safety is a common topic for any hvac/r group, whether it's awareness at a construction site or routine building maintenance. Introducing a new employee to the firm and maximizing the time needed to train individuals are two ways audiocassettes and videotapes can save time and money in any company.

TRAVEL

Managers should always plan for delays when traveling. If you do a great deal of flying, postponement of your flight can catch you off guard. Waiting for your airplane to depart can leave you with an hour or more to sit and do nothing. Your strategy for handling this dilemma should be to not only plan on having delays, but also plan on arriving **early** at the airport terminal. I find it less stressful to arrive with plenty of time to spare. Long before I arrive at the airport, I have already planned what I will be working on while I wait for the plane. This planning will offer you the opportunity to work on a specific project or report. Because you have planned ahead, you can intentionally put off scheduling time to work on this specific project or report until you have reached the airport. Once at the terminal, you can perform what I call **laptop work**. Some work must be completed in the office, but some work can be performed later at the airport terminal. By looking ahead, a manager can schedule what work should be done, where it will be done, and when it will be done. Planning allows you the opportunity to concentrate on essential elements while in the office and still meet your deadlines for those reports that can be completed outside the office.

I also plan what I will work on in my hotel room after I have arrived at my destination. Scheduling work time for later in the evening is another opportunity to maximize those hours you are out of the office and away from home. In your room, you have comfortable surroundings and the benefit of an electrical outlet for your computer. Sometimes it is cost effective to have room service bring your dinner to you rather than spend the time eating alone in a restaurant. This idea may seem excessive, but I consider it time well spent. When I return home, the work is done and I have time for my family. Years ago, I would do my writing using a pad of paper and a pencil. When I returned to the office, my secretary hated to see me because she would have stacks of handwritten data to type. The administrative secretary encouraged me to start using a laptop computer. Now she doesn't dread seeing me walk in the door after my trips.

A laptop computer is a great time management tool, and I take mine with me everywhere I go. In addition, I always carry in my briefcase the necessary tools for designing an hvac/r system, so I won't have to go back to the office. These tools include a Trane Company "ductulator," pipe sizing chart, architectural scale, psychrometric chart, and a standardized concept package, which lists the numerous components necessary in a construction-managed hvac/r system (a sample concept package is included in Appendix A). With my computer and the contents of my briefcase, I am able to complete most of my work at the jobsite. People often travel long distances to reach a jobsite and then realize they don't have all the tools to perform the job. Again, planning ahead can eliminate this dilemma. Bringing the necessary instruments, allowing ample time to complete the task, and engineering or surveying the site carefully will maximize an individual's time away from the office.

If you are driving, not flying, to a jobsite, I recommend you complete your work there rather than returning to the office that day. When you reach your destination, allow enough hours in the day to complete the task at the site. If need be, arrive the night before or stay another day to complete the work. Prepare a list of things to do, and monitor your progress. Whatever you do, don't go back to the office until you have completed all the work you set out to do. Returning to the workplace, when the work could have been completed at the site, will most likely result in inefficient use of your time. People asking questions and telephone messages are just a couple of the distractions you can expect when you walk in the door. The site work you were going to complete at your desk is almost always delayed; and the longer it's delayed, the harder it is to remember the details which were so fresh in your mind at the site. When you return to the office to complete these final details, something important frequently is missed and valuable hours are wasted. Returning to the site a second time will be costly, both in man-hours and travel expenses. Recognizing the optimum time to perform the work and using that time effectively are basic elements of good travel time management.

The time you spend at the airport and in your hotel room can be used effectively to catch up on your work. With no telephone calls or people walking into your office, this travel time planning can provide you with valuable hours to write. End-of-the-month reports, engineering project reports, documenting estimate historical data, and developing corporate training lesson plans are just a few of the many tasks that can be completed with a laptop computer away from the office and away from home.

Managers who know they are going to be out of the office can better plan their in-office workload. Many writing tasks can be completed when a manager is waiting for a plane to leave; while assignments requiring teamwork need to be accomplished in the office through communication with other workers. Planning allows managers to evaluate their workload options. Knowing you need to complete a report and not anticipating a delay at the airport, may not be good planning on your part. Then again, if your workload is light, reading the paper in the terminal may be a relaxing alternative. Travel time management and the effective use of time management tools will aid a manager and other workers in the efficient use of their days and nights.

BUSINESS MEETINGS

How you control a meeting is fundamental to good time management. Very often, employees will comment on the number of meetings that occur in the workplace. These comments usually have a negative connotation when the remark is made. As a manager, it is important to the group that meetings emphasize a **positive** process by a collection of contributing staff. Effective use of a person's time begins with the directions given by supervisors. Through good management, employees can learn how to work better.

Actions speak louder than words! Nothing could be more true when it comes to managing other people. Demonstrating and advocating effective use of a person's time will have a contagious effect on the entire workforce. A manager needs to make it obvious to the workers that each person's time is important to the success of the company. A productive **meeting agenda** and the implementation of the company's **meeting ground rules,** will benefit both managers and employees (meeting ground rules are discussed in further detail in Chapter 4). Managers need a successful game plan to demonstrate how they want to see a process implemented. The success of a meeting begins with an action plan that includes the agenda and the meeting guidelines. Employees need to know ahead of time that the manager is planning on this meeting being a worthwhile use of their time. At the end of the meeting, everyone should appreciate that their time was well spent, because everyone followed the game plan.

Prior to the meeting being scheduled, the person calling the meeting must appreciate that it will take individuals away from their other tasks. Therefore, this meeting must have a purpose and a goal. The person calling the meeting should coordinate the time with an agenda that people also can appreciate. The length of the meeting, when it will start, and when it will end are very important factors. You should never schedule a meeting unless you indicate the commitment of time needed to assess the issues. The person arranging the meeting is also responsible for maintaining the meeting notes. A useful method for planning the meeting notes is to expand the agenda issues into individual topics, which can be modified by the author as the topic is discussed. The meeting notes should be identified as an **action agenda**, Figure 1-1. This action agenda should be a standardized sheet with the following pertinent data:

- Project name and job number (when applicable)
- Meeting number and date of the meeting
- Name of the person taking the meeting notes
- Names of the attendees
- The names of the people who will receive the meeting notes should also be recorded. This will include the attendees, as well as other people who should be kept informed.
- Each **action item** should have a number. Preferably, the numbering should be a combination of two numerical values. The first value should be the meeting number, and the second value should be the action item introduced at this specific meeting, i.e., 1-4 means the fourth item on the agenda for meeting number one.
- Each action item should include a due date or deadline and the name of the person responsible for completion of the task.

By putting it down in writing, there is no mistaking who will be responsible for the completion of each task. The meeting notes should always end with a statement requesting anyone in attendance to please respond by a predetermined time period if there is an error in the accuracy of the meeting notes. No response is a confirmation that the information is correct and that the individuals responsible for completing an action item will meet their commitment.

If the meeting is held outside the office, a manager should recommend to the author of the meeting notes that the notes be completed prior to returning to the office. Issuing meeting notes in an expedient time period is critical to individuals meeting their action item deadlines. If the notes aren't issued in an appropriate period of time, commitments will slide and the time spent meeting and discussing the importance of these issues is lost. If the author of the meeting notes chooses to return to the office with the notes incomplete, these action items tend not to be completed or issued in an acceptable time period. This usually occurs because the

---

Action Agenda

Project: _____   Job #: _____
Issued by: _____   Date: _____
Attendees: _____   Distribution: _____

_____   _____
_____   _____
_____   _____

Action Items   **Responsibility Deadline**

1-1 _____   _____ _____
1-2 _____   _____ _____
1-3 _____   _____ _____
1-4 _____   _____ _____
1-5 _____   _____ _____
1-6 _____   _____ _____
1-7 _____   _____ _____
1-8 _____   _____ _____
1-9 _____   _____ _____

Sheet __ of __

Figure 1-1. Action agenda.

writer is distracted with other commitments back at the office. The meeting is an important event requiring the attendance of numerous individuals, but it becomes somewhat less important when the meeting notes are issued late. A manager should recognize that this tardy action negatively affects those who attended the meeting. Remember, this assembly of personnel took these individuals away from their own priorities and responsibilities in order to attend that meeting.

Just as important as the action agenda is controlling the time spent in the meeting. Regulating the flow of the meeting is a difficult task that must be performed if time management is going to be successful. The person scheduling the meeting should start on time, and those who are late for the meeting should not be allowed to waste the time of those who are punctual. Starting on time sets a standard and gives a signal to those attending that the person calling the meeting is in control. During the meeting, conversations should be kept to the business at hand. Side conversations should be discouraged, because they detract from the purpose of meeting as a group. Finally, ending the meeting on time is just as important as starting on time. People in attendance will usually have other commitments on their calendar that will require them to leave at the prescribed time. If the issues haven't been satisfactorily addressed, then another meeting should be scheduled with only those workers necessary for this second meeting.

SPECIFIC TIME MANAGEMENT TECHNIQUES

There are many different techniques you can use to effectively manage your time. Having already discussed those techniques pertaining to training, travel, and business meetings, I would now like to discuss some additional techniques and standardized forms, which I have found to be very helpful in managing time efficiently.

"Things to Do" Lists

Remember the moment and write it down! Almost every day I think of a new idea, something important I need to do, or a person I need to talk to. These thoughts and ideas are usually instinctive and leave my head almost as fast as they enter. This may not be a unique occurrence to some, but I would venture to guess that most, if not all, people in responsible positions have very active minds. The difference between the select few and the many is that the select few get into the habit of **writing down** their random thoughts and ideas. When I first started out in the working world, I was taught to develop a **"things to do" list**, Figure 1-2. This invaluable tool has served me well over the years.

Every employee should be taught to keep a "things to do" list for their daily, weekly, and monthly activities. I prefer to keep my notes on a pocket-sized sheet or pad, which I can keep with me when I leave my work area. Usually this list is kept on my desk, and I will look at it first thing in the morning, just before lunch, and at least one more time before I go home at night. I realize this is a simple concept, but it helps immensely with prioritizing and planning activities and commitments. A good example of assigning priorities is the identifying of travel time work and laptop work. Depending on the due dates, this work can be scheduled for an appropriate time when you will be out of the office and/or traveling. Keeping a "things to do" list gives a person the opportunity to schedule the time needed to meet his or her deadlines.

Priority	Task	Due Date
A, B, or **C**	Bring latest technical magazines to read.	April 1
A, **B**, or C	Write next month's technology column.	March 25
A, B, or C	Review Jim's final draft of hospital report.	March 3
A, B, or C	Call ahead for drawings to be available.	March 1
A, B, or C	Call ahead for testing and balancing report.	March 1
A, B, or C	Bring camera, flash, two rolls of film, and batteries.	March 3
A, **B**, or C	Arrange lunch with Facility Engineer first day at site.	March 4
A, B, or **C**	Update the department goals for the quarterly review.	March 21

Figure 1-2. Example of a "things to do" list.

Time management experts suggest using letter or number designations to rank the priority of the items on the list. In our example, we have used letter designations. "A" is the highest priority, "B" the next highest, and "C" the third level of urgency. In all cases, these tasks are important, otherwise they should not be on the list. This technique can be helpful to most people when it comes to prioritizing their time. Another method is to designate the due date after the priority. This inventory of chores should always be on one page. If the list extends to another sheet, I find the tasks on this second page tend to be overlooked or forgotten. Another problem with two or more sheets is that tasks tend to be too general. The tasks must be specific and brief. If the list continues to be too long, then it is possible the person isn't getting the job done. The list will become shorter as tasks are completed.

As a manager, you should share this strategy with your employees. Highlight the benefits of keeping a task list, and teach your employees how to maintain one. A manager should continuously monitor each person's list for responsibilities, compliance, and completion. Monitoring the use of "things to do" lists by individuals in a group is a useful method for tracking their progress and commitment to the job. More importantly, both the manager and the employees should keep these lists and use them for their own personal efficiency!

Note Taking

Keeping a note pad in your car is another way to catch a creative moment. If you listen to audiocassettes, there often will be a valuable point of interest that you want to remember. A note pad next to you in the car offers you that opportunity to scribble it down. "Scribble" is the correct word to describe how I jot down my notes in the car when I'm driving. I have perfected the art of scribbling on a pad of paper without looking at it. In addition to information obtained from tapes, you may be thinking of what has to be done as you drive to work. It seems like your subconscious mind functions at the same speed at which you are driving, and random thoughts and creative ideas can come and go in a flash. I try to catch these thoughts and ideas. If you don't write them down while they are fresh in your mind, these ideas may be gone.

Whenever I go into a meeting, I always bring my weekly/monthly calendar and a separate note pad. The calendar is for scheduling any action items that may result from this or a follow-up meeting. The small note pad is for jotting down random thoughts and ideas that may pass through my head while I'm attending this conference. This is particularly useful if the meeting is not moving very efficiently and time is being wasted. I can fill in the void periods with thoughts of what I need to do after the meeting. In addition, I may have an hvac/r problem on my mind, and I will use this time to think the issue through while the meeting drags on.

Something as simple as a pad of paper can be a valuable time management tool. You should also get in the habit of carrying a "things to do" list with you wherever you go. Keep one on your desk and another in your car. In addition, get in the habit of bringing a pad with you to meetings to fill in the voids that can occur. These note pads need to be coordinated for information and prioritized for deadlines. Reviewing the responsibilities should be done on a daily basis. When the task is completed, the idea should be removed from or crossed off the list. A by-product of using these lists is the satisfaction you get when you have accomplished your tasks. It is rewarding to be able to remove one more responsibility from the priority list. Keeping it up-to-date and documenting deadlines are important to the success of certain action items. Also, make sure the list is no more than one page and pocket-sized, so you can keep it with you. The items you record on this inventory listing should be those which are most important.

Non-Think Work Lists

Another useful time management technique is the identification and effective implementation of what I call **non-think work**. I first became aware of this production task when I was overseeing an hvac/r design that required a significant amount of drafting. I identified this job as non-think work because it was very repetitive. A manager has to be creative sometimes when work begins to back-up. On this particular project, I had to complete 38 drawings, all of which included a number of non-think tasks. These tasks included drafting in the room names and numbers, column letters and numbers, title blocks, and drawing titles, just to mention a few. This was also before we had computer-aided drafting and design (CAD). A person didn't have to be an engineer or a designer to complete the work. In fact, the person didn't even need to be a draftsperson. So I recruited a few of the secretaries to work overtime and showed them how to use a template to draft the names and numbers onto the drawings. The work required approximately five minutes of explanation and took about eight hours to complete. The secretaries did the work on overtime, which was something they weren't used to doing. The work was a nice change for them and was necessary in order to finish the drawings, so they performed the work with a lot of pride. Even more importantly, the assignment freed up the experienced drafting staff to concentrate on the more technical drafting chores.

Non-think work can be described as any task that takes minimum time to explain and is very repetitive. On most projects, I have the lead engineer develop a list of non-think tasks at the beginning of a job. During the completion of construction drawings, these tasks serve as good fill-in work. A guideline for this non-think work is that a person must be able to spend less than five minutes explaining the task. This way, if people finish up an assignment and have half an hour or more before they are scheduled to begin another task, they can pick up a non-think job. On another project, I had 98 construction drawings to complete, and this technique of identifying non-think work was invaluable when it came to keeping people busy. Figure 1-3 is an example of the type of non-think work list I often use.

Non-Think Work List		
Task	____% Complete (Shade box)	Remarks
Drawing title		
Title block		
Company logo		
North arrow		
Key plan		
Shade key plan		
Legend		
General notes		
Engineer's stamp — signed		
Room numbers		
Room name		
Column letters		
Column numbers		
Flow arrows		
Fire dampers and shafts		

Figure 1-3. Example of a non-think work list.

Another example of non-think work is the mail you receive each day. When my mail is delivered, I look at it immediately and sort out what needs to be read that day. The rest of the mail I will drop on the floor next to my briefcase. At the end of the day, when I pick up my briefcase, I will pick up this mail and bring it home with me. At some time during the evening, I will sort out and read this "less important" correspondence. You should not waste time during the day reading data that was not on your agenda for that day. At home, it usually takes only a few minutes to determine if the mail is of any value. If it is a magazine that I'm interested in reading, then bringing it home offers me a chance to read it with fewer disruptions. At the same time, I will quickly go through the magazine and tear out only that information I want to read.

The mail you receive should be checked first thing in the morning, just before lunch, and just before you leave for the night. Depending on the hvac/r business a person is involved with, a manager should monitor what needs immediate response and what can be scheduled for another time. By checking your mail routinely, you can control your time while controlling priorities.

Managers must be cognizant of the value of both their time and their employees' time. Often, the work required to complete a job can be performed with little direction. A manager needs to maximize the talents of each individual in the group. At the same time, the manager must be prepared to fill the workload voids, no matter how brief these periods may be. Maintaining a non-think work category in the office or at a jobsite allows a manager to achieve these goals with little effort. Employees will also appreciate the value of this method and practice this process as team members in a team effort.

YOUR PERSONAL WORK ENVIRONMENT

Desk management is just another way of saying time management. I have always been envious of the person with a job that allows them the luxury of a clean desk! As a manager with 17 years of management experience, I have yet to find that elusive profession that allows me to function with a clean desk. Instead, I have learned to manage my desk. The following few paragraphs describe the desk and office arrangements that I have found to work the best for me. Do not feel you must set up your personal work environment in exactly the same way as described. These are simply guidelines showing one way of managing your desk, so you can make the most efficient use of your time as possible.

The first priority in desk management is to have your most important work in front of you. This is the work you are currently working on and plan to complete first. In addition, your calendar should be out in front of you, and it should be open to the current week. The daily calendar lists the meetings and commitments scheduled for each particular day. In my daily calendar, I like to place any upcoming memoranda or agendas into the specific week where they will occur. This way, when I turn to that week, the correspondence I have placed there ahead of time will provide me with information I will need regarding upcoming events and issues. Putting the correspondence into the calendar also helps to ensure that it isn't misplaced.

At the upper left-hand corner of my desk is my mail basket. I keep it here so I can routinely check (three times a day) the incoming correspondence and daily mail. At the upper right-hand corner of the desk is my next work priority. Its placement serves as a visual reminder of my next task. Should an unexpected task come in, I can quickly assess its urgency and rank on my list of priorities.

The next place I look to file active projects is to the right of me, on an adjacent table. Here I can place the work folders, files, and correspondence that will require my attention in the coming days. Each stack needs to be neat and orderly. It is also in plain view to serve as a visual reminder and for convenience should there be a need for the data. To the left of the desk, on the floor, I drop all my evening mail and any correspondence that needs to be sent out. This outgoing mail may be internal correspondence or mail to be sent outside the office. When I get up from my desk, I make a point of picking up this packet and dropping it off at my secretary's desk for distribution. Sometimes it is necessary to hand deliver the information, but sometimes placing it in the outgoing mail basket is sufficient. It is a handy way to keep the data out of sight and out of mind while I'm working at my desk. At the same time, you can't miss these documents when you leave your desk because you have to step over them.

Directly behind me is the telephone. That too is out of sight and out of mind. Telephone interruptions can be difficult to control. If a person has the luxury of a personal secretary, then the secretary can screen your calls. A good set of guidelines is needed to make this work, because the hvac/r industry is a service business. If your firm is going to continue to be successful, customer service must come first. Together, you and your secretary can develop a priority list to control the telephone calls that get through to you. The secretary must know which people are customers and which people are vendors. You may also want to limit telephone disruptions during the times when you are at peak performance. I consider myself to be a morning person and believe that is the optimum time when I get the most work completed. As a result, minimizing the interruptions during the early part of the day allows me to complete many tasks that can't be delegated. You may also have to provide special instructions on a day-to-day basis depending on what you are working on and/or your commitments for that day. Telephone management is integral to managing your time. When the phone rings, it is essentially asking you to stop what you are doing to address someone else's agenda. Developing a **telephone priority list** can streamline this activity to meet your customers needs and allow you to fulfill your commitments, Figure 1-4.

Following this simple strategy will increase your productivity while at your desk. Within most offices, people have a limited area where they must juggle workload, telephone interruptions, and incoming and outgoing correspondence; accommodate routine meetings; and present a professional and well-organized space. Plan your workspace to accommodate the business needs of the firm, and devise a simple game plan to handle the flow of work as it passes over your desk. At the same time, add a personal touch to the office that exhibits your outside interests and allows people entering the room to relate to your individuality.

	Yes	No		Yes	No
Customer #1	___	___	Vendor #1	___	___
Customer #2	___	___	Vendor #2	___	___
Customer #3	___	___	Vendor #3	___	___
Customer #4	___	___	Vendor #4	___	___

Hold calls 7:00 a.m. to 9:30 a.m.
(Monday, Wednesday, Friday)   [Yes][No]

Hold calls 7:00 a.m. to 10:30 a.m.
(Tuesday and Thursday)   [Yes][No]

Current project activities/critical milestones/commitments:
    Monday: _____
    Tuesday: _____
    Wednesday: _____
    Thursday: _____
    Friday: _____

Figure 1-4. Telephone priority list.

SUMMARY

Always remember that your time is important. Be in control of your life! Time management is the cornerstone to realizing self control with almost everything you accomplish. Maximizing each hour of the workday will complement your efforts to succeed as a manager.

In order to be a good manager, you must be very good at managing your time and the time of those who work with you. This begins by acknowledging the period of the day that you are at peak performance. At the same time, encourage your staff to be conscious of their peak performance periods. This is the period of the workday when interruptions should be kept to a minimum. Organize your work schedule around this interval of time, so you can increase your work output and the output of those working with you. Train the entire staff in all areas of time management, so they can continually strive for peak performance.

Help create a corporate library that will assist in the continuing education process of each employee. Each year budget a nominal amount of money to invest in audiocassettes and videotapes. In addition, produce in-house training tapes applicable to your company's business. Keep this inventory on a master list that is updated quarterly, and share the list with your customers. Yours is not the only firm struggling with this educational endeavor and you won't be the first to establish a training program or company library, but you can be the best at educational awareness!

Tutor members of the company in travel time management. This tool offers workers the opportunity to improve the quality of their work and helps them use their time efficiently. Using your time effectively is integral to meeting deadlines and being profitable. Advocate the use of laptop computers, which are invaluable devices for increasing a worker's performance, both in and out of the office. Today's technology also offers each of us the possibility of filling workday voids with productive work by planning ahead. This planning can maximize the time you spend outside of the office and allow you to return to work without a backlog of chores that could have been completed while you were gone.

Using a few rules and guidelines, you can complete your work by scheduling the appropriate time to perform these tasks. When scheduling tasks, use a non-think list detailing those items that can be completed by individuals with little direction and time. At your desk, maintain a "things to do" list. Prioritize the items if necessary and keep the list brief. Finally, don't go anywhere without a note pad. Keep a pad in your car, and bring one with you to meetings. When you leave your desk, take a pocket-sized pad with you. In fact, **don't leave the office without it!**

Chapter 2
Making the Change from Employee to Manager

When I graduated from high school I went on to attend a trade school, where I took a one-year course in drafting. When I was finishing high school, I really wasn't interested in going to college. At that time, continuing my education would have been a waste of my time and my parent's money. Instead, I wanted to pursue my education by learning to be a draftsperson, which would also allow me to start earning money. Based on the drafting needs during that time period, the school was able to ensure job placement upon graduation. At the end of the school year, I accepted a job in the hvac/r industry in a consulting engineering firm.

In the northeast at that time, many hvac/r engineers were not graduate engineers. These people had entered the business through the trade sector working for contractors, service companies, and installers of hvac/r systems and components. Others were graduate engineers with limited exposure to the "real world" of hvac/r. Unlike the theoretical information engineering students learned in college, the consulting engineering business has often been described as **cookbook engineering**, and learning the business has always been an on-the-job experience.

At my first job, I was fortunate enough to work in an environment that offered me the opportunity to learn. My first employer had only six other employees besides myself, and each was very helpful with my on-the-job training. Within six months I was sizing ductwork and piping. The work was challenging, and I enjoyed having the opportunity to draw the systems and coordinate their placement in the ceiling space. By the year's end, I was drawing the hvac/r system layouts for various equipment rooms and coordinating their accessibility.

THE 80/20 RULE

Growing up within the business, I was introduced to the philosophy that 80% of the people in management positions are *good* at their jobs but not necessarily *great* at what they do. This is called the 80/20 rule. These positions included project engineers, project foremen, service technicians, and managers. If a person understood the basics of the job and had a lot of common sense, then he or she could reach this 80% level. Many hvac/r design engineers without college degrees were able to advance by following a design process based on these two simple premises: **this is how you do it** and **it worked this way the last time**. Hvac/r system design was simple in the '60s. Systems were often oversized, because bigger was thought to be better and wasting energy was not a significant concern to most designers. I became an excellent example of the 80/20 rule, because I was considered a project engineer only four years after I had started in the business.

As I progressed through the ranks of the company, I moved from drafting to design engineering to engineering. Eventually I became the project manager and was responsible for coordinating the contract documents and specifications with the other in-house engineering trades and the electrical and plumbing designs. This was my first opportunity to manage people beyond my hvac/r design team. During my last three years at this firm, I was the project manager and hvac/r engineer on every project I was involved with. At the same time, I volunteered for and was assigned the additional job of developing the company's master specification and standard details. I did this other job on my own, at home. Although I didn't know it at the time, I was developing an appreciation for the benefits of standardization and the benefits of knowing the details. As a manager, you will always have to be attentive to details. These details may be related to your personnel, project, or finances. Throughout your management career you will continuously need to focus on the details. For me, this was a good beginning.

I also learned valuable information when compiling the data for the master specification. This was a great opportunity to learn all of the ins and outs of the hvac/r business. Whether you are a manager for a consulting engineering firm, construction company, service company, or building maintenance operation, you need to know the specifics of your field. For example, in order to properly oversee a facility's operation, a facility manager must know the equipment, preventive maintenance techniques, the impact of hazardous waste, associated codes and regulations, etc. The best way to meet this demand is to be proficient with the details and specifications associated with each component.

This extra work helped me temporarily supplement the theoretical engineering information I lacked having not attended college. It also helped me separate myself from the crowd. My first employer once told me "when you stick your head above the crowd, you leave yourself open to get hit with a brick!" This backlash can be expected from a minority of fellow workers, but it shouldn't distract you from striving to be the best you can be. It is this attention to details that has helped me pull away from the 80% and join the 20% who are truly great at what they do.

During my first eight years at this firm, I had the benefit of working on a regular basis with two valuable engineers, Jim McGrath and Hank Eggert. Hank and I were a team. We were usually given the building renovation projects to design instead of new building projects. Over a period of time, I learned to appreciate the value of working on renovation projects. It was imperative that you paid attention to details and, more specifically, the existing conditions. Hank was good at details, and by working with him I improved my skills and became a more thorough design engineer. He also showed me tremendous patience, a skill I have yet to master. In those first eight years, I never saw him show his frustration with others. Managers must show similar patience if they are going to get the most out of each individual they are responsible for. Seldom will a worker respond in a positive manner after being treated harshly. It is your task to provide constructive criticism in a manner that will be accepted by the other person. Shouting will never deliver the message you are trying to send.

Jim McGrath, the Chief Engineer, was the person who hired me. He always took time to show his interest in me and assist in my education. Jim was a graduate engineer and had worked for a construction company before crossing over to the consulting business. Fundamentally, his engineering knowledge was sound, because he knew the theoretical aspects of hvac/r engineering. In addition, he had spent a number of years working and gaining practical experience in the construction business. It was Jim who introduced me to the use of a "things to do" list and the development of check sheets to control my projects.

I have reflected back on these first few years of hvac/r training many times, and I now know that a person needs to strive constantly to excel. Very few managers possess that unique ability to "stick their head above the crowd." A manager should learn to pay attention to details **and** see the full scope of the situation. At the same time, workers need to have direction and learn from the manager's experience and knowledge. A lot of workers will "talk a good story," but few will strive continually to put forth the extra effort needed to succeed. This isn't to say that people don't work hard to get ahead in business. You can go to any evening class to see how many employees are working to improve themselves. The workers I'm talking about are those who think they could manage the company better than the existing management staff. These critics will question the policies and procedures of the company in idle conversations but seldom contribute to improving the firm. To excel in management, you constantly need to strive to improve the workplace for both the company and the employees. This takes persistence on the part of the manager, each and every day; and when you are tired of being the standard bearer, step aside and let someone else carry on. Otherwise, you will fall back into the 80% group of managers who don't excel at what they do.

EDUCATION

Hvac/r managers must learn the fundamentals related to their profession! Today, more than ever before, a college education is essential to achieving advancement within the engineering segment of the hvac/r industry. At one point in my career, I worked for an architectural, structural, mechanical, and electrical consulting firm. It was at this firm that it became clear to me that I had to go to college if I was to get ahead. During approximately four years at this firm, I went from being a senior engineer to eventually managing the hvac/r staff. When the hvac/r manager left, the company didn't replace him. Instead, I was given the responsibility of managing the staff of four while continuing to be responsible for my own engineering projects.

At that time, the hvac/r industry was beginning to experience dramatic changes. The 1973-74 oil embargo had introduced the beginning of the energy crisis, and hvac/r technology responded with drastic changes in the way we could design various systems. Also, the days of working your way up through the ranks without a formal education were diminishing. Future engineers will need to be college educated to fully comprehend the sophisticated systems created in response to the energy crisis and our conservation awareness. Because these world-wide changes were affecting the hvac/r industry, I became well aware of the limitations being presented to me by the company.

In response to the changing technology, I went to college in the evenings to learn the theories behind my practical knowledge. A manager doesn't have to know everything about hvac/r; however, he or she must be **fundamentally sound** in order to make the correct decisions and cannot work by the philosophy "it worked this way the last time." In this business, graduate engineers may not be able to fully utilize their academic skills in the first few years. However, five or six years down the road, when they are in a responsible position, they will make design decisions based on fundamentally correct knowledge. Theoretical experience and practical experience are two important elements that are so vital to being a good manager. In addition, I set for myself the personal goal of becoming a registered engineer, because professional credentials are also essential to effective management.

In 1976 I returned to my first employer. An interesting observation I immediately made note of upon my return was that I wasn't "the kid" anymore. Since that time, I have noticed that it is not unusual for a person to start at a firm right out of high school or college and never shake the image they presented at their entry-level position. A manager needs to be cognizant of an individual's progress. Within a week of my return, the boss called me into his office to express how impressed he was with what I had learned since I had left the firm. I don't think he fully recognized the skills I possessed when I originally worked for him. Although he assigned me projects to engineer and manage when I first worked for the firm, I still had that image of a kid new to the business. Upon my return, I looked older and was older than many of the other employees. I wasn't the youngest person on the staff anymore.

Managers must never overlook the skills available within the company, particularly those workers who have successfully progressed from within. Often, a firm will hire people from outside the company to fill management positions, because they failed to recognize the in-house skills of those employees who have worked their way up through the corporate ranks. When looking to fill a management role, one- and two-year training programs can be established for on-staff candidates, rather than seeking an unknown applicant. At the same time, the selection of an existing employee to fill a management position must be based on that person's ability to satisfy the job requirements. Companies will frequently promote people because they are loyal, long-time employees, not because they meet the position description criteria. A manager should not be promoted from within unless he or she meets the qualifications set by the company. This is a common error and one that is difficult to correct once it has been made. In the end, management is not satisfied with the individual's performance, and the individual becomes disenchanted with the company.

A hurdle some newly promoted managers may face is the issue of youth. I was at least ten years younger than the next youngest manager at one firm. This was a new environment for me to be working in, and I was responsible for approximately 38 people, including engineers who had more than twice the years of experience I had. In addition, I was still attending night school and was yet to be a registered professional engineer, whereas some of the people I was responsible for were graduate engineers and registered professional engineers. In a discussion I had with my boss about this concern, he wisely advised me to **do my job**. If you are going to manage and you believe you are the correct person for the job, then do what is needed to succeed. In other words, he was saying that actions speak louder than words. When I moved into management, I realized that I didn't have time to convince everyone that I was in charge. I had to get on with business. This was important advice for me as a new manager, because it is almost impossible not to feel like you have to convince your staff that you are qualified to manage. If you are in control, your employees will recognize and accept you in that position.

Young managers need to recognize that hvac/r is not rocket science. It is an environmental technology that requires technical experience. A person working in the hvac/r industry needs the engineering fundamentals that are taught in college, but common sense is also integral to achieving technological success. Most hvac/r systems are a collection of off-the-shelf components. Engineers aren't creating the first mouse trap, they are just striving to build a better one! As the manager, you don't need to know all the answers. However, you do need to have a fundamentally sound grasp of engineering theory; you must know where to look for answers; you must be a stickler for details; and you must be a good sounding board for the design engineer's questions. If you don't have college training in mechanical and electrical theory, then you will truly have trouble being effective in today's hvac/r industry.

CHANGING JOBS

I eventually left my first employer in an attempt to expand my on-the-job education. This move was very difficult, because I liked the company, the people, the work, and the opportunities. But I felt like I was becoming too comfortable in my position and that this was not good for me or the company. This is an important observation that a manager needs to make with themselves and the people for whom they are responsible. People who become comfortable or complacent in their positions need to know the impact of this on their futures. Often, employees will talk about "getting ahead" but are not consciously putting in the extra time needed to educationally advance. A manager needs to make note of this fact when discussing the person's performance. There is nothing wrong with not getting ahead, as long as the person is aware of the ramifications of the status quo! Not everyone is expected to move to the top. Managers need to understand the types of people they are exposed to in the hvac/r industry and the types of people they will be responsible for managing. In my case, I wanted to grow educationally, and it was important for me to see how other consulting firms did business.

I have known others in the business who wanted to make a move but hesitated. As the manager, you should be cognizant of these individuals. Employees who are contemplating a job move should be able to discuss this move with you. It may be in the best interest of both of you to talk and agree on a course of action. There is nothing wrong with leaving a company for the right reasons. For example, I recall one worker who wanted to make more money than we considered him to be worth. What someone earns can have a negative effect on the day-to-day performance of that person. If you cannot agree on an acceptable salary, then the person should move on.

At the same time, I have known individuals who stayed with a firm even though they were not satisfied with their jobs. For example, I knew a man who was at the peak earning level for his job position. He became locked into the job because no other company would hire him due to the excessively high salary he earned. It is not unusual for someone to "earn" his way up the financial ladder of success into a position where he is the highest paid person for that job. This usually occurs to someone who is just on the outer edge of management and hasn't been able to cross over into that arena of activity. This becomes a frustrating situation for employees, because they can't go any further within their present employment, and yet they are locked into a salary they don't want to give up. It is your job to direct them into being satisfied with their present position or to encourage them to look elsewhere. In either case, it is up the worker to do something about this dilemma.

The firm I started working at when I changed employers had a staff that I would categorize as average and in the 80% zone. I lasted three months. The firm wasn't aggressive enough in seeking new work. This company operated at a very comfortable pace requiring a nominal 40 hour work week. I recognized that I was not ready for this and preferred to work additional hours. I was used to working by the philosophy "if you can manage two jobs, we will give you three jobs." One benefit of this work ethic was that my education was accelerated because of my added responsibilities. Workers who are anxious to grow in the business should be concerned with their long-range growth potential when working at a comfortable pace.

A manager must be cognizant of the **Type A** worker who enjoys and seeks out the opportunity to do more. This worker shouldn't be confused with the person who wants as many hours as possible just to earn more money. Both of these individuals have a different agenda. In my case, I wanted to learn more. Money was important to me, but it was not the reason I took on the added work. At the same time, there are workers who are comfortable doing additional work that is repetitive, and they will do this work to earn more money. As a future manager, this experience gave me the opportunity to add these personality types to my catalog of people in the hvac/r industry. Knowing the different types of individuals who make up this business will help you as a manager when you are working with or are responsible for these employees.

Another lesson I learned from working at this second firm was that it is good to work for a bad company. This is an important revelation for any employee and manager. Often, people think "the grass is always greener" at another company. Managers should remember and appreciate the talent and experience of past employers while working with their present firms. If you are employed at a company where you hate to get up and go to work in the morning, recognize the issues that make this firm a "bad company" and remember this experience. No company has cornered the market on the perfect environment, and it is important to appreciate the benefits of a good firm.

It is important to a manager to be open-minded to the deficiencies that can exist within a company. Often, when a person becomes a manager, that individual will strive to correct all the employee complaints. This can be anticipated by upper management, because the new manager has just crossed the street from employee to employer. A manager must look out for *both* sides of the street — management and employees. However, you should acknowledge the fact that management cannot solve all the problems or keep all the employees happy all the time. A manager must be fair *and* realistic to maintain a smoothly running company.

When I left my second job, I had the opportunity to return to my first employer or move on to a different company. Because I had not yet achieved my goal of learning how other firms handled consulting engineering, I went with the latter. This new firm specialized in the architectural and engineering design of industrial facilities, as well as industrial automatic control system design. Both of these areas of engineering were new to me. My experience up to that point had been in health care, office buildings, and educational design applications. The new firm also worked with general contractors on construction management projects. This was a significantly different method of engineering than the "plan and spec" method I was familiar with.

The construction management concept offers managers the opportunity to be more creative with their management skills. The interesting challenge of construction management is that the engineered product must be completed at a much faster pace. In order to accomplish this, equipment is pre-purchased, and sheet metal and piping installations usually begin before the final architectural floor plan drawings are completed. Also, a great deal of teamwork is needed to finish these projects on schedule. This new work opportunity offered me the chance to broaden my education.

Individuals who are anxious to learn more about the hvac/r industry should recognize the benefits of a job change and what it can offer in the area of continuing education. Employees and managers should continually strive to increase their knowledge. For the manager, being in charge doesn't mean the education process can slow down or come to a stop. Hvac/r technology continues to change with each new year. However, you shouldn't make a change if you are doing it only for the money! I like to observe how new managers arrange their offices when joining a firm. If they focus on how "executive" their offices look, then I become suspicious of their priorities and reasons for joining the company.

Did they come to learn and teach, or did they come for the salary incentives and image?

Usually, companies will pay individuals an appropriate salary for the positions they are attempting to fill. A good rule of thumb is to be paid within ±10% of the salary commensurate with the job responsibilities. It is the individual's responsibility to keep abreast of the "market value" for the position description. A good way of doing this is to communicate with other people in the business. As a manager, staying in touch with other managers can help you determine a fair value for your employees. The by-product of changing jobs will usually be more money with each move, but the opportunities that the next employer offers should be the focus of your interest. Successful, long-term managers are those who have worked their way up through the employee ranks, changing employers as needed and striving to be the best they can be, for themselves and for the people they manage.

CREATIVITY — A MANAGEMENT BUILDING BLOCK

The person who can combine organizational skills with creative expertise can be a great manager. Those people who are in charge of others must be well organized if they are to be successful in fulfilling the job description of a manager. Creativity, on the other hand, is seldom considered when choosing a new manager. However, this unique talent can be invaluable to the person in charge. Corporate inspiration can keep employees involved and keenly interested in what they are doing. Workers appreciate a change of pace in their daily routines, and it is the manager's responsibility to be the inspiration in a creative environment and a positive influence for the team. This person's ingenuity can make the difference between good work ethics and great work ethics. As a manager, you should constantly strive to maintain a creative climate in the workplace. Creativity and challenge within their routines result in an upbeat atmosphere for workers.

When starting out as a manager, creativity can often be misinterpreted, because the new leader is inclined to bring change with his or her new responsibilities. Imagination and creativity need to continue over the duration of your career, not just the first year or two. It has been my observation that managers lose that ingredient as they become comfortable in their role of boss. These seasoned veterans inadvertently lose touch with their workers as the years pass. They forget what it was like when they were on the other side of the street. Managers need to renew their "commitment to lead" each year and should begin each day with a creative and organized mind. People don't automatically renew their position description annually, but it is a good habit to get into. In future years, the quality process will offer managers who have this attribute the opportunity to use these tools to the fullest. The quest to improve through the quality process complements the individual who may have just joined management, as well as those experienced managers.

Employee participation and open communication can also assist the manager who lacks a creative mind. Through teamwork, management can make changes that will give the company a fresh outlook on business. The quest for perfection complements new managers, whether they have an enterprising attitude or are seeking this characteristic. Either way, creativity can be a management building block.

Finally, creativity is a unique consideration when selecting a manager. Any person who is in charge of others should also be cognizant of this skill. Imagination generates excitement and the desire to get up and come to work each day. This tool can do more than most other management methods to inspire job excitement and employee satisfaction with the work they perform. First-time managers need to draw on this tool as they take charge and introduce the transition with excitement. At the same time, creativity can prove to be the difference in securing a project contract, solving a system problem, and/or satisfying a customer's need. This distinctive tool can make the person a leader among leaders. If individuals work on this talent, they can use it as a building block to better management.

BE PREPARED

Whether you are moving toward a management position or are currently in a management position, always be prepared! A manager needs to be able to anticipate the workload, the competition, and the next move. When anticipating the next move, you should constantly be striving to do better and to do it before someone suggests it to you first. Plan your meeting agenda, the project scope, the field site visit, etc., a day or two in advance. This gives you ample time to think through the agenda and prepare the others who will also be in attendance. By distributing the agenda, everyone will have a chance to contribute. At the same time, this pre-planning demonstrates your commitment to the issues.

It's always good to receive input from others, because you are a team. This advice can be from more experienced managers, or it can be from other workers. Any good advice is well worth listening to. As the leader, it is important that you continually offer the ideas and/or direction. After all, if you are not constantly planning the departmental goals and direction, then who is? Finding ways to improve the company, employee satisfaction, and project processes is an important aspect of your job. Don't get wrapped up in your own individual job tasks and forget to lead.

Workload maintenance is a good example of anticipation and planning. The most cost-effective team is usually the smallest team. How you succeed at this strategy is dependent upon how well you and the members of your group are prepared. This is particularly important with a business that has large peaks and valleys. The service business can be like this with its seasonal start-ups and shutdowns. A good manager must be proficient in scheduling the workload to meet client needs. At the same time, you don't want to hire more staff and then have to let them go at a later date. By gathering the group together, you should be able to resolve this annual issue without compromising the experience the team has to offer. One possibility is to persuade the employees to buy into a work schedule that will pay them a weekly salary based on forty hours work while performing more than forty hours labor. Months later, when the workload has dropped off, the employees still receive their forty hours pay while working less than forty hours. If everyone understands that it is in the best interest of the company *and* the workers, then these people will do their part to make it work.

Being prepared for the competition is always a challenge. As the manager, you should participate in the sales and marketing efforts of the company. This may be in a support role, but your experience can offer a significant advantage to the company. This advantage can be a benefit when comparing how your firm performs versus the competition's performance. Whether the business is a consulting, construction, or service firm, the business standards are the same. For example, with construction firms, competing companies will approach the project in a similar fashion. Each will follow the standard procedures necessary to build the system, process the paperwork, pay the bills, and close-out the job. The benefits you can bring to the project are your individuality, unique experience, and reputation. How these are presented is dependent upon how well your team is prepared. By continually striving to be the best manager, you can bring these benefits to your clients through your efforts to be the best at what you do.

SUMMARY

When making the change from employee to manager, it is your responsibility to continue to excel. Because you are now a manager doesn't mean you can just sit back and delegate. Managers must still be anxious to learn, ask questions, and build reputations as the best at their profession. This should be a continuation of the reputation that allowed them to be promoted to management in the first place. Once you reach the management level, your weekly workload increases dramatically. Now you are responsible not only for yourself, but to management and the workers. Making the change just made your workday longer and hopefully more rewarding.

When making the move into management, the new member may experience a change in the attitude from the other employees. A person who makes the transition from worker to manager can't let other people's opinions interfere with his or her commitment to manage. Assuming the new person in charge wanted this position and worked to earn it, **getting the job done** is necessary for success. Remembering that 80% of managers are good but not great at what they do, new managers have the chance to break free from the crowd and become a part of the elite 20%. Moving into the high performer category will take time, literally years! But the opportunity is there if the manager is willing to take it.

As a new manager, a person probably won't become too comfortable in the new position very quickly. However, over time this person does need to be aware of this occurrence. A leader must be out front leading! Success often takes the excitement out of daily routines. As time passes, the new manager becomes a veteran at the job and those issues that were once important to this person don't seem as meaningful. I have seen some bosses who have forgotten what it was like on the other side of the street. The manager needs to set the pace through example and through motivation. When leading the team, concentration on the details can be an effective use of one's time. Establishment of checklists and the standardization of processes can greatly assist a first-time manager once he or she has taken charge.

In addition to providing leadership, managers must be sensitive to the individuals who make up the team. As a person makes the transition into management, employee requirements are pertinent to the overall success of the company. By being promoted, new managers have reached one of their goals. Remembering that other individuals have goals and desires, these new bosses must divide their own time between company goals and those needs of the workers. You can't forget that it is a two-way street when managing! In order to build a better team, a new leader must get the most out of each worker for their individual benefit, as well as the benefit of the company. This means coaching each employee to strive for success and advancement. When making the change into management, you should strive to stick your head above the crowd. Watch out for the brick though! Be different than all the others, and do it with confidence and commitment. In order to do all this, you must prepare yourself for the next day, week, month, season, and year. You must also prepare your staff to continue to improve while you cultivate the skills needed to excel in management.

Chapter 3
Tools and Techniques for the First-Time Manager

If your goal is to be a manager, then you are going to need a plan! When I was chosen to be the manager of hvac/r engineering at my company, I had approximately one month to prepare my plan. At the time of my promotion, I had not anticipated my boss making a change in the management structure. However, I did have a strong opinion as to how the company was being managed, and this proved to be a hidden blessing. For the first-time manager who is also new to the company, this may prove to be a problem. The new, inexperienced manager will probably not know enough about the issues at hand to grasp the solutions. First-time managers need to be careful when applying for the job of manager if they have not been in the firm for a sufficient period of time. Your management "game plan" may not be adequate for the job. If a person has been in the company for a sufficient period of time, then he or she should be prepared to implement a well-orchestrated strategy to enhance the operation. If you don't have this experience, then you have to ask yourself "what can I offer the company that distinguishes me from the others?"

THE THREE-YEAR PLAN

With only a month to collect my thoughts and back up my management opinions, I began to put my management game plan together. On my own, I sat down and began to formulate my two-way street philosophy. I put down in writing how I thought a manager should manage, and I welcomed the opportunity and challenge to be a valuable asset to the company and its employees. You shouldn't go into a management position thinking you are the smartest person in the group. Instead, you need to be the most determined person in the group in order to improve performance. Your convictions, which are based on past experiences, observations, and opinions, should be sufficient to be a successful leader.

After I was given some direction from management on "how to be in charge," I formulated my own plans. Fortunately, I had earlier training in the use of a "things to do" list, which proved to be the vehicle to organize my thoughts. I also used this logical method to set goals. My strategy and goal was to develop and implement a **three-year plan**:

- **Year One**: Put in place (establish standards of how business was performed)
- **Year Two**: Make it happen (my plan in action)
- **Year Three**: Repeat the success (prove the plan can continue to work)

I cannot say I shared this philosophy and strategy with my boss. It wasn't that I didn't want him to know what I was doing, I just didn't think to present my plan to him due to my inexperience. In the future, when taking on a new management responsibility, I always put down in writing my plan, goals, and the necessary time frame and present it to my boss.

Year One

Year One of my plan was simply to "put in place" my methods of management. I don't think drastic changes are needed unless the company's problems are at a critical stage and you have had a long time to think about the solutions. In my case, the changes were scheduled to be subtle, and the course of action was to be implemented at my discretion. Year One was primarily to identify the standards of how business was performed.

A manager should methodically catalog all aspects of the product produced or service performed by the company. By numbering each task, issue, or category required to achieve the end product, you should be able to separate these tasks into "stand alone" goals. Each goal becomes a target that will be measured during the first year for compliance. An example of this is to note that standard details are used frequently, but these details have not been updated in the last five years. In this instance, the goal is to not only update the standard details with the latest technological advancements, but to incorporate them onto computer software. You should indicate milestones that are keyed into specific dates during the year, such as the developing and indexing of details to be reviewed during the coming year; reviewing the details for shut-off valves; drafting the details using a computer, etc. As these dates come and go, you can check the progress of your workers and their ability to fulfill the goal on schedule. As manager, you should post these goals in a location where everyone can follow their progress.

The standards of how business was performed should also address employee needs. Business has to be a two-way street! A manager is required to represent the employees, as well as the company. One of my goals was to establish a consistent salary structure within the group for which I was responsible. I listed the workers by the extent of their hvac/r knowledge, separating them into job classification groups such as draftspersons or designers. This **position status chart**, which is discussed in further detail in Chapter 8, let people know exactly where I considered their skill levels to be. If I was right, the person had the opportunity to work towards the next position level. If the individual disagreed with me, this listing offered that person the chance to meet and discuss his or her opinion regarding the position skill level. In addition, I also established salary ranges for the different categories. Although individual salaries were not posted, the salary ranges were known. A manager should make sure, annually, that the people within the group are compensated fairly for their work. This listing, established as a Year One goal, solved that dilemma.

Year Two

Year Two of my three-year plan was devoted to "making it happen." For a manager, putting the plan into action is only the beginning. During Year One, the goals were identified and the tasks documented in order to improve the department. Year Two needed to demonstrate the plans or goals used, the concepts implemented, and the success achieved. For a first-time manager, existing data may not be available to compare and measure the success of the new plans. In subsequent years, I learned that measurement is essential to proving the results. Collecting past data to measure future activities can be an arduous task, but it should be completed to give credibility to the process. As a manager, I noted the changes as I made them. This method allowed me to document what was changed and why. In subsequent years, these notes provided past performance information that proved useful to me, detailing what worked and what didn't.

The second year needs to be successful from a business point of view, as well as with the employees. Job satisfaction can be contagious, and a good manager needs to be the leader in spreading this enthusiasm. With my three-year plan in place and everyone aware of its existence, active participation by other employees added to the success of the goals. Position listings encouraged the workers to join in and help, because they knew the company had taken the time to recognize their value within the firm. The existing staff knew the ground rules regarding the plan. The workers knew they were integral parts of the company and understood the philosophy of the two-way street. These factors are very important if a business is to grow and succeed and products or services are to improve.

Year Three

My goal in the third year of management was to repeat the success we had in the second year. I needed to be able to show the company that we could achieve financial growth and increase our repeat business that year. Likewise, a company needs to show continued commitment to its employees. As a result, our customer base remained solid, our reputation grew through work on various prestigious hvac/r projects, and the average number of years an employee stayed with the company grew by approximately 50%.

I believe my success as a first-time manager was directly related to my past training. This training included time management skills, organizational skills, goal setting, and note taking. As a result, the transition was a success. From management's standpoint, I would recommend that all new managers participate in a training program to ensure they are capable of managing others and will succeed at the job.

When the candidate is ready to take charge, the new manager should have a three-year plan similar to the one I outlined earlier. This plan should be well documented with goals, time frames, assumptions, objectives, and results that can all be measured. The new manager should present this plan to his or her boss for approval. The boss should offer suggestions and opinions based on past experience but should always sign-off on the new manager's plan. New thoughts, plans, and actions can impart new life into any progressive company.

Your three-year plan should be created to establish a new direction for the company, as slight a change in course as it may seem. You should maintain a set of standards detailing the proper methods for conducting business, as well as a set of personal standards. This lets every worker know what is expected from business performance and business results. At the same time, these standards should ensure each employee that they will be treated fairly, or in other words, on a level playing field. The second year is the year for scheduled results. This is the period when you should be putting your plan into action and fine tuning it as needed. The third year is the time to repeat your success, reinforce your management skills, and prepare to expand the process for creating improved profits, customer satisfaction, and employee satisfaction.

SETTING THE PACE

As the person in charge, the manager must always **set the pace**. If the work schedule starts at 8:00 a.m., then you had better be working before that time. You should always remember to **manage by example**. Too often managers send out a signal implying the "do as I say and not as I do" philosophy of management. The manager must always be out in front leading the charge. It is unrealistic to expect the staff to set the pace while you are in the rear trying to keep up with your own work. Some managers will be sporadic with their take-charge spirit, and some will lose interest in being the leader as they become comfortable in their jobs. If the manager doesn't continually strive for proficient operation, employee satisfaction, and cost-effective results, then that person's leadership and authority will begin to be challenged. When this happens, the employees are going to start to question management's effectiveness, the direction of the company, and their own job satisfaction.

Prior to my promotion to manager, I was an assistant manager for the hvac/r group at my company. I performed these responsibilities for approximately one year. During that period, I believe I worked aggressively to improve the engineering effort within the company. As I did my job, there was very little management feedback regarding my efforts. At the same time, I was disenchanted with the performance of a few people in the hvac/r group, because they were not keeping to the pace I wanted to see them achieve. Because I wasn't the manager, this area of the hvac/r operation was beyond my position of responsibility. The responsibility of encouraging these workers to do better, work harder, and be cost effective still rested with the hvac/r manager.

Some of the employees were comfortable in their positions and were not performing at the same level as other employees. There was perceived to be a double standard for performance among the employees. This double standard is not acceptable. A successful team must know that everyone is doing the best job possible, and if they are not, they must know that management will address the problem. If a worker's performance is not satisfactory, it is the manager's responsibility to bring this problem to that individual's attention. I'm not saying that no one can have an "off day," but you must get up the next day and do your best.

I learned this "setting the pace" work ethic from my father, who worked six days a week selling cars. He had a commitment to his family and to his boss to do his job. Growing up, there was an unspoken rule in my home: **work hard and do what you are supposed to do**. My father set an example without ever telling me about the long hours he worked. The job of a manager takes up more than 40 hours a week, and if you are not prepared to put in the necessary hours, then you are not going to be able to set the pace for your workers. Right or wrong, some people will always look to your faults. Managers have to be leaders all the time, and they must be consistent with their leadership and strive to have each employee work just as hard.

Setting the pace is fundamental to managing. A manager must instill in the group a desire to work hard, which shouldn't be confused with "hard work," such as lifting eight-inch steel pipe! My understanding of working hard is consistent, continuous use of your time while doing the job right. At the same time, a manager needs to do more than look busy. Employees recognize a sincere, hard working performance by their boss. This performance will also be contagious. I have seen employees work efficiently as team members, while the less motivated employees tend to perform randomly at a similar rate, so as to not look out of place. If a manager does not lead continually, then the less motivated employees are going to look for those opportunities when they can slack off. These same people are going to use the manager's lackluster performance to justify why they are not putting forth the extra effort. It is amazing how often disgruntled workers will blame their poor performance on someone else.

ACCEPTING THE NEW POSITION

When I was promoted to the position of manager of hvac/r engineering, I received a nominal salary increase. I wasn't really concerned with the financial aspects of the job, because what was important to me was to be responsible, in charge, and successful. I truly believed that I could make a difference, both to the company and to the individuals within the company. This is an important statement, because many managers only strive to be in positions of authority so they can earn more money and/or increase their status. This leads us to another fundamental rule of management, which is **do the job first and then get paid**. I want to earn as much money as I can just like everyone else, but the key word here is **earn**. I have never accepted a new position within a company that stipulated a salary increase with the position change. If I am successful in the new role, I expect to be compensated for my performance but not paid for it up front. If you are a first-time manager and truly want to achieve company improvements, focus on those improvements first. Also remember that these improvements need to be measurable and accomplished within a reasonable time frame.

If you are overly concerned with a fair value for your services, then your priorities are not in line with those of a first-time manager. If you let this salary issue influence your everyday attitude, then you are the wrong person for the job. The number one responsibility of a first-time manager is to get results! If you can't contribute management performance improvements, then you didn't deserve the opportunity to perform. Some new managers may disagree with this opinion, but I believe everyone has a responsibility to **prove** they can do the job, whether it is as a draftsperson, designer, technician, engineer, project manager, or department manager. Anyone can say they could do a job better; but talk is cheap, and a good manager should avoid this temptation.

In fairness to the employer, new managers should recognize that they have done nothing at the management level to justify a salary commensurate with experienced managers. In addition, this financial issue should have no place on your list of priorities. First-time managers should focus on implementing all the steps, changes, and measures that ignited their need to manage. It is beneficial for new managers to recognize this philosophy, because they will be confronted with this same attitude as workers within the group advance and change positions. As a manager, you will be confronted with this "pay me first" attitude more than once. If you accepted your promotion with the attitude that you wanted the opportunity to perform first, then it is easier for you to support this opinion. If your salary was a very important factor in your decision to accept the promotion, you have established a double standard that other employees may or may not know. I know I can look a person in the eyes and truly express my philosophy, because I've been there.

CROSSING THE STREET

An observation I made prior to my crossing the street into management was that employee criticism towards the company should always end on a positive note. What I mean by this statement is that a person has the right to complain, but they should also be prepared to offer a solution to the problem. When I was promoted to the manager of hvac/r engineering, there were a couple of other staff changes. One of these changes included a promotion for another person into management. This new manager had been very anti-management prior to his promotion. Seldom did he offer constructive criticism; instead he just complained. When he crossed the street into management, he changed his outlook on the company. So drastic was his attitude change that his fellow employees became skeptical of him and his leadership. He had made a 180 degree turn, and in the process lost a considerable amount of credibility from the other workers.

When making the change to a management position, you need to take your integrity with you. It is true that a new manager must make adjustments in his or her new position, but a subtle transformation is needed. People need to know that you are going to work to implement changes expressed prior to your promotion. Many times I have seen managers replace their past reputations with new images. One such person transformed from a "blue collar" image to the suave, sophisticated businessman image. His effectiveness as a manager was badly damaged from the beginning. Former co-workers resented the transformed technician-to-manager image he portrayed.

People who are promoted often inadvertently change who they are. With each new position comes the challenge to perform and succeed. Those who don't enhance the job criteria inadvertently reinforce the **Peter Principle**, reaching the point of complete incompetence. In other words, some people advance up the corporate ladder because they are doing a good job; but with each promotion, they take on new and different responsibilities. Eventually, they reach a position where they are not qualified to manage. Each job description will have a protocol, and you need to follow those guidelines. However, nothing happens overnight. Transformation to your new position should be subtle. Present yourself as a new leader who is committed to both sides of the street — employee and employer. Try to be yourself in the process. People will value your efforts and support your goals if you leave your ego behind.

When preparing to move into a new position, a person needs to be working on his or her policies and procedures long before receiving the promotion. By example, a potential manager can lay the building blocks that will someday form the foundation of how he or she will manage in the future. This approach will help you prepare for the change and avoid the dilemma some leaders have when they take charge. I often reflect back on my experience, attitude, and work ethic before I became a manager. In addition, I try to remember what my financial status was at that time when discussing cost benefits with employees. A manager needs to contemplate the logic behind the employee's opinion. You were on the side of the workers before you were promoted. It is important that they know you value their opinions and that you haven't forgotten them now that you are a manager.

DRESS CODE AND THE OFFICE ENVIRONMENT

When accepting any promotion there are certain added responsibilities that will affect how you present yourself. This may include a **dress code** change, whether the advancement is from technician to supervisor or from project manager to department head. The new position may require a white shirt instead of a blue shirt or a suit instead of jeans. Changing the dress code doesn't mean a person has to change his or her commitments, opinions, values, and/or personality. Instead, this is an opportunity to expand your values and ideas. You are now representing the company, and maintaining professionalism is essential. If possible, a future leader should be dressing for the position they want, not the position they have.

Certain areas of the workforce do not have the option of "dressing up." If a worker is a technician, he may be required to wear the uniform given for that job. However, you can strive to be the best dressed technician with polished shoes, a cleaned and pressed uniform, and a well-organized tool box. This daily routine sends a message to clients and your fellow workers that you are a professional. Along with this daily announcement, employees must enhance their presence through performance. It is important to be "recognized" as the manager, but success will speak for you as well.

Learn what a professional appearance means to the company, and work to pass these same standards on to the other employees. An experienced manager may have forgotten what it was like to operate from an employee's smaller weekly budget. At the same time, employees don't automatically think to invest in expensive clothes, because the boss wants them to look like managers. By communicating with the workforce, a manager can encourage the employees to consider the value of a dress code.

It is also important that management doesn't overreact regarding dress codes. For example, I had the opportunity to observe the dress code policy at one company, which recommended each male manager always wear white long sleeve shirts. They didn't have a dress code for female managers, because they didn't have any at the time. I could never understand the logic of this policy. It seemed to me that the company had more pressing issues to address than to spend any time discussing this preference. These superficial policies give management a bad name. Surprisingly, some managers become preoccupied with these silly issues rather than focusing 100% on the business plan.

Other issues managers should be aware of include making coffee and keeping the office environment presentable. Everyone should take the time to join in when it comes to the workplace. Employees always make note of those who drink coffee but are apparently too valuable to take the time to refill the pot. A responsible manager with good time management skills always finds time to participate in daily office activities. In fact, it can be a refreshing break from your daily work! A manager should encourage people to enhance their workspaces with certificates of achievement, plaques, pictures, etc., that personalize their area and send a sensitive signal that these employees care. In the common areas of the office, encourage the showcasing of company achievements, such as project photos, letters of appreciation, awards, newspaper and magazine articles describing accomplishments, etc. Clearly, a picture is worth a thousand words, if not more.

It is management's responsibility to encourage the employees to take a keen interest in their work environment. This is their home away from home. How this area is maintained is an important topic. A cluttered, dirty, or bland space can affect the work atmosphere. A manager needs to encourage the other employees to share in a spirited environment. I'm usually the first person into work, so I take on the responsibility of watering the plants in the office. Plants add color to a dull workplace. At one company I worked at, we had a spring cleaning day. Once a year the company allowed everyone to come to work in jeans and T-shirts. Everyone was provided with dust cloths, cleaning products, and plenty of trash bags. At noon we would have a cookout in the parking lot. The Chief Executive Officer and I did all the cooking. It was a nice change and a great way of cleaning house annually.

SETTING GOALS

Having been brought up keeping a "things to do" list, I naturally progressed into **goal setting**. To my surprise, the skill of goal setting, which I describe as building a management road map, is a tool not often used by many managers in the hvac/r industry. However, a new awareness in the last few years has made managers work to improve their goal setting abilities. To some, this skill doesn't come easily, but because of my earlier training, this skill comes naturally for me and is the first thing I think to do when given a new responsibility.

Setting goals can be compared to taking an automobile trip from Boston to New York. Drivers automatically pull out a map and review how they are going to get from one city to the other. The same strategy applies when managing a group, department, project, or company. The person in charge needs to outline a "road map" that documents how they are going to achieve their goals. For each goal a manager sets, he or she should list the following elements:

- The goal
- The theme
- The known issues, topics, facts
- The milestones and time frame necessary to complete the goal

The **goal** should be brief and precise regarding what is to be achieved. A manager can make certain assumptions regarding the **known issues**, such as recognizing the procedures that are in place and the obstacles and/or deficiencies that exist. Additional time will be needed to confirm these assumptions, but this can be a good start. More importantly, documenting the objective, or **theme**, is paramount to achieving the goal. The manager then needs to spread the goal activities over the preselected time frame. These goal activities are **milestones** along the way to reaching the goal. Each individual should also have company-related goals to work towards during the year. These delegated goals can offer employees an opportunity to participate in the success of the firm and can be a measure of their performance during the year. These delegated goals should also be part of employee annual reviews and a requirement for any salary increases.

Referring back to the example of the road map from Boston to New York, the manager can demonstrate this process to the employees, Figure 3-1, so they can use this example to establish their goals.

By outlining your plan and the time frame necessary to achieve this plan, Figure 3-2, a manager can routinely monitor the progress of each and every goal that is set. Achieving success relies on the **measurement** of the goal. Has the project been started? Is it 50% complete? Is the goal on schedule, and will it be completed on time? All these questions can be quickly checked and the responses confirmed by reviewing the progress on a calendar.

When I was chosen to lead the hvac/r group at my company, I was prepared, because I had been trained in goal setting early on in my career. This skill is an essential building block to managing people in the hvac/r industry. Understanding the process of goal setting takes time, just as it takes time to experience the benefits. Once a manager feels comfortable with this management tool, he or she should expand this responsibility to the employees. Delegating this administrative tool to others can relieve you of some paperwork, while helping other workers develop their own goal setting skills. By putting down in writing your first-year plan (using the method above), a new manager can be in control from the beginning and stay on course throughout the year, proving the results at the end

	Road Map Example	Hvac/r Example
What is the **goal?**	New York	Development of a master specification
What is the **theme?**	Drive a car to New York City	There is no master specification at this time. The spec should be computer based using word processing software.
What are the **known assumptions?**	Drive carefully and at posted speeds	Begin with an index. Have the spec cover all hvac/r equipment, material, subcontractors, general conditions, and installation.
What are the **milestones** along the way?	Interstate route numbers	Selection of computer software. Assign responsibilities to the personnel working on the goal. Assign time to complete the tasks.
What are the **time frames** needed to reach each milestones and the goal itself?	Odometer mileage and time	Dates when index, spec boiler plate, general conditions, equipment, etc., will be completed.

Figure 3-1. Example of "road map" strategy for setting goals.

of the year. The same can be said for those employees who assist in the group's annual progress by using the goal setting tools described earlier.

1995 Goal: Develop a master specification **No. 1**

Theme: There is no master specification. A computerized word processing software package should be developed.

Known issues:
- Develop an index
- Select a software package for word processing
- Complete the master spec in three segments (general, equipment, and material)
- Format must follow industry standards

Milestones:
- Develop an index that includes boiler plate general conditions (complete by 1/15)
- Develop equipment list and subcontractor scope (complete by 3/1)
- Review computer software (complete by 2/1)
- Select the software (complete by 3/1)
- Complete general conditions (complete by 5/1)
- Complete equipment specifications (complete by 9/15)
- Complete subcontractor specifications (complete by 12/15)
- Use Division 15600 Series

Jan.	Feb.	March	April	May	June	July	Aug.	Sept.	Oct.	Nov.	Dec.
1/15	2/1	3/1		5/1				9/15			12/15

Figure 3-2. Example of how to outline a goal and the time frame necessary to achieve the goal.

Summary

Long before a person is promoted to a manager's position, that individual needs to set the foundation for his or her management plan. Ideally, you hope that management is in tune with the person's skills, talent, perception, and leadership. Unfortunately, upper management will probably not recognizes these attributes, and the prospective manager will need to formulate a plan to overcome this situation.

Employees who believe they can do the job better must prove themselves through their performance. They need to back up their words with positive results. First-time manager candidates can set the pace long before the position has been offered to them. You have to plan ahead **before** you are given the responsibility to lead. My opinion and experience is that a three-year plan is adequate time to demonstrate that you will be valuable to the company in a management role. To reiterate, the three-year plan is as follows:

- Year One: Set the standards of performance
- Year Two: Demonstrate on a daily basis that your management tools will enhance the operation
- Year Three: Repeat the success and improve the deficiencies

When you accept a position as manager, focus on the job at hand. Don't make your salary the first order of business. Do the job first, then get compensated for your efforts. When "crossing the street," recognize that it is a two-way street. You need to look out for both the company and the individual workers. A new manager needs to prove that the firm has a product worthy enough to invest in, namely you! If you are as good as you believe you are, then the company shouldn't have a problem reimbursing you when there is a measure of performance.

Leading up to the promotion, the candidate should already be dressing for the position. "Free spirit" dress attire is not appropriate. The hvac/r industry is a profession that has various dress code requirements. Engineers and project managers should wear business-like sports jackets, suits, or outfits; technicians should wear company furnished pants and shirts; and managers should wear somewhat more formal outfits or suits. Dressing appropriately is important, but remember that actions speak louder than anything you might be wearing. Dress codes don't prove you are a professional, they simply complement your performance. If a manager, project manager, and/or foreman does the job efficiently, on schedule, and within budget and the client is satisfied with the results, then people will recognize their value and abilities.

And finally, put your first-year plan down in writing. Identify the objectives, assumptions, milestones, and time schedule for each goal you plan to achieve in your first year of management. While working towards the achievement of these milestones, train your staff to assist in the process. Through selective designation of job responsibilities, you can delegate these chores effectively and train the associated workers in the process.

Chapter 4
Maximize Your Leadership

We have all heard the saying, "actions speak louder than words!" This is simple advice but so very important to the hvac/r manager. I believe that if managers are to lead, then they must be able to do so through their performance. A manager should let his or her actions do the talking. An example of this is getting people to be prompt at the start of the day. Employees don't want to work in an environment where "do as I say and not as I do" is the prevailing attitude of management. If you expect people to start work on time each day, then you had better be there working when it's time to start. Set the pace and your workers will follow. Efficient use of your time can be contagious. Co-workers are quick to assess a manager's performance. If the person in charge is a conscientious worker, then most of the employees will "buy in" to that kind of work environment.

Working my way up through the ranks, I became aware of the influence of good work ethics. Starting as a trainee, I observed the actions of others in the workplace. I noticed the effects of both positive and negative actions. Listening and watching those people who were in charge was an extension of the work ethics I learned from my father. Like many of the "blue collar" workers of his time, he labored for long hours, six days a week. I inherited my work ethic from him, and in the process, my attitude and philosophy towards management and the need to lead was formed. By working hard, making efficient use of my time, and doing more than my share of the tasks, I believe my actions have spoken for me. Whether you are a new manager or have been in the position for a number of years, a leader must lead! You can never let up when it comes to setting the pace.

At the same time, a sense of humor is also essential if you are to lead effectively. Deadlines must be met almost every day of the year, which in turn causes a great deal of stress. Keeping the staff focused requires periodic "light hearted" breaks throughout the workday. If people are going to enjoy the constant pressure necessary to achieve their goals, they must do so with a little humor. Creating an environment that has workers wanting to go to work and feeling good about what they do is a responsibility the manager shares with each of these employees. Keeping that environment healthy is the responsibility of the person in charge. Don't delegate these obligations to anyone but yourself. You can recruit assistance with these tasks, but don't relinquish them completely. Remember, a leader must lead!

Don't Toot Your Own Horn

I once had a conversation with a fellow employee, who told me how good he was at his work. Another worker overhearing the discussion noted that "you shouldn't toot your own horn." I doubt the person doing all the talking heeded this advice, but I always remember the comment. It had a hidden message, which told me that **if you were truly good at what you do, then people will take note**. In addition, they will do the talking for you. This is particularly true in the hvac/r industry, because word travels quickly when it comes to skillful personnel. Good people are hard to find, and the hvac/r industry is no exception.

Employees often take up the habit of highlighting their own achievements when management doesn't take the time to notice a job well done. This fixation is due to a couple of different reasons. First, they may never have worked in an environment where people received a "one minute praising." To the best of my knowledge, the book *The One Minute Manager*,[1] written by Kenneth Blanchard, Ph.D. and Spencer Johnson, M.D., introduced awareness to "one minute praising" as a management tool. You will find that a brief word of encouragement or a "job well done" is beneficial when working with others. I have used this tool in the past, long before Mr. Blanchard and Mr. Johnson were good enough to give it a name. Now, I find it easier to remember to use this tool for communicating my support to an employee. If the workplace is an atmosphere of frequent "one minute praising," then the process could become contagious, and the tooting of horns will diminish. Frequently commenting on a person's performance is integral to the employer-employee relationship. A worker wants to know that his or her performance is appreciated and that it is the result of positive actions.

Another reason workers tend to speak up for themselves is because of their own success awareness. These employees are apt to thrive on compliments and wear praise like a badge of honor. This type of person may be difficult to bring "down to earth," and you will find them reluctant to accept constructive criticism when the time comes. I have also noticed that this kind of person will be outspoken against management. Many times, these comments are a smoke screen used to divert the manager's constructive criticism away from themselves. This is particularly noticeable when the person offers an "off-the-cuff" solution to a problem. Anyone can throw out a quick answer, but most problems require a well-planned resolution. If the issue has been around a while, it is a very good possibility that the solution requires a team effort to come up with an amicable settlement.

I can recall at least half a dozen employees who missed the opportunities made available from constructive criticism. Instead, they would highlight their past accomplishments and/or experience as a rebuttal to management's comments. Each time, they would toot their own horn. They wanted and had a need to justify their actions by diverting the attention away from themselves and toward another issue. Sometimes you can't get through to these kinds of people, but their co-workers are very apt to recognize the problem and the corrective measure. At least you can seek comfort in knowing that the other workers are learning to do better from your message.

Work performance that results in a benefit to the company and/or the individual is a healthy situation that should be encouraged by management and supported by the employees. A manager should also encourage other people in the firm to take the time to speak up for a job well done. Also, recognize those individuals who praise their own work and try to get them to focus on satisfying the work at hand and not individual accomplishments.

My Donut Story

One morning I was in the Financial Manager's office when she commented to me about a worker who had not yet started to work even though it was 8:45 a.m. I recognized that she was very good at her job as financial manager and that she was committed to working extra hours, if necessary, in order to meet her responsibilities. There was no question that she was one of the most responsible employees in the firm, but she sometimes wasn't sensitive to other workers' points of view.

On this particular morning, she was aggravated by this other person's informal discussion of a football game the night before, on company time. I pointed out to her that while we were both discussing a "business issue," the other person could see that she had a cup of coffee and a donut on her desk. From the other worker's point of view, she was having a coffee break and was not working while she was talking to me. A manager should recognize that other employees will assess what the manager "appears" to be doing, so they can justify their own actions. Some workers don't really care if you start fifteen minutes early or that you work an extra hour three days a week. There are people who do a good job and do it during the normal working hours. When 5:00 p.m. comes, they don't care if you work through the night — they did their job and now it's time to go home. For managers to sit back and think that everyone recognizes and appreciates their work ethic is a mistake.

I have often heard comments made to other workers by the person in charge. The manager making the commentary may think the person is not working as hard or as efficiently as the manager. At the same time, these workers may be unaware of the remarks being directed at them. There are usually two sides to an argument, and there can be two sides to how a person is performing. As a manager, you need to be very observant and anticipate the points of view of others before offering constructive criticism or a reprimand. You may be surprised, if not astounded, by the worker's response to your remarks.

When talking to other employees about leadership, try to encourage them to always set the pace. If you are going to lead a team, workforce, or department, you need to be out front leading the charge. There is nothing more discouraging than to have the boss in the rear shouting "charge." When criticizing a person's performance, a manager needs to assess all the facts and evidence. Even if you think you are setting the pace, you should still assess your own presence as a manager. Are you projecting the "do as I say and not as I do" double standard? Employees can be quick to point out management's errors and a sub-par performance. They may not bring these observations to the manager's attention, but they will certainly discuss their observations with their co-workers. It also doesn't help when slow workers can justify their actions by pointing out that, although they may not be working, you are sitting there eating a donut and not working either!

DON'T SHOW YOUR FRUSTRATION

This may be an unreasonable request, but a manager shouldn't openly express frustration to a group of people when the issue is with one person. Like the "one minute praising" management tool discussed earlier in the chapter, there is also the "one minute reprimand."[2] The ability of a manager to speak to workers regarding something they have done wrong can be a difficult task for the manager to perform. However, it helps if you hold the attitude toward this dilemma that constructive criticism is in the best interest of the individual with the problem. At the same time, reprimanding workers in front of other employees is not the correct thing to do! When giving advice, a manager should do so one-on-one. It does no one any good to suppress the topic, and if the worker is to grow in the business, then it is in his or her best interest to know what's on your mind. If the problem encompasses a group, then take the group aside and offer the "one minute reprimand."

While offering direction, always emphasize the positive by using the term "we." Employees need to know that long-term success can be achieved only through teamwork. Individuals should recognize that they are in a responsible position that requires an accountability, but they are not alone. When I first became a project manager, the company's chief engineer, Jim McGrath, instilled in me the "we" concept. We were employees and representatives of the company, having been given this authority by the company. Nowhere did the word "I" and "my" have a place in this philosophy. I was taught to use the word "we," as well as the phrase "work with" instead of "work for."

I believe the potential for frustration can be reduced if there is a team attitude. A manager can address a project/personnel problem much more easily if the other individual believes the instructions are being given to improve the situation. Employees who talk in terms of "I," "my," and "works for me" have more difficulty accepting this constructive criticism. Also difficult is the manager's dilemma when working with this type of person.

Another frustration managers experience is when a lead engineer, project manager, and/or foreman makes a significant error in judgment or operation. These are the people who managers count on daily. When an error is made, it is never made intentionally, but the damage is done. It has been my experience that many managers confront this issue differently. An experienced and confident manager will have no problem assuming "co-accountability" for the problem. In this case, the worker can be removed from the "hot seat," and the manager accepts the responsibility. If these problems become a common occurrence, then the manager will have another frustration — continuous inadequate employee performance! A less experienced manager or a manager who doesn't like to be wrong, will react differently when a problem arises. These types of people will either not support the person who made the error or present themselves as understanding managers. In either case, these managers will distance themselves from the lead engineer, project manager, and/or foreman, because they have a problem accepting total responsibility and being accountable for another person's actions.

Frustration can also be perceived by employees as a sign of management weakness. In addition, it can be discouraging to the employees if the manager outwardly expresses frustration on a regular basis. Disappointments can be a daily occurrence if the manager has not created a qualified group of employees, and these occurrences can distract the team from having a positive attitude. It is the leader's responsibility to continually reinforce the goal to do better. Using the "one minute reprimand" and sharing the responsibility when something goes wrong are tools of a good manager.

Workers should accept any error they may make and learn from their mistakes, while keeping these errors to a minimum. Likewise, workers should not feel that if they err, they will be terminated. Accountability is essential to leadership, and this leadership needs to be delegated to the workers. Managers cannot do everything, so they must share some decision-making tasks with the employees if progress is to be achieved. If you hire and train good people, then frustration can be kept at a minimum.

IMAGINATION

An attribute many managers don't have and probably will never have is **imagination**. Some people are born with a creative mind; some will work to develop this skill; and some will never see the merit of it. In order to rise out of that average 80% and into the elite 20%, a leader needs to be enterprising. Originality is often perceived by fellow workers as a positive action that provides results. These results speak for themselves when a project is successful. The creative manager who initiates a better way of doing something inherently demonstrates a unique skill that can complement leadership skills. If you are a very creative manager and skilled in time management, then others will recognize your guidance and direction. This comes in very handy when encouraging people to work harder, more efficiently, and more cost effectively.

Even before I was a manager, I would often ask "why not," because sometimes the obvious isn't obvious. A manager should always be open to new ideas, without trying to reinvent the wheel of course. You should encourage others to question the process, concept, or practice. Hvac/r fundamentals haven't really changed over the years, and engineering basics are still the same. How we design, install, maintain, and service the hvac/r systems and equipment of yesterday, today, and tomorrow is where we can excel. Every company wants to be successful and profitable. The difference between one hvac/r firm and all the other local competition may be its imagination. Taking a different point of view can separate a company and its individuals from the rest of the crowd. Hvac/r technology was built on this concept.

A good way to involve other workers in this enterprising atmosphere is to frequently gather the team together for a **brainstorming session**. Collectively, your team will be able to provide a better engineered product, a better installation, and/or a more efficient hvac/r system. The by-product of these sessions is that everyone can learn from the process, and each person will build a better understanding and hopefully a better appreciation of the individuals who work together. These team "think" meetings need to be managed effectively to achieve the intent of the brainstorming session. The manager should outline the ground rules at the beginning of the meeting, including those found in Figure 4-1.

A manager should also inject a sense of humor into the discussions. As the leader, you can demonstrate to the others that any idea may have a value no matter how crazy it may seem to the others. A humorous alternative may be worthy of serious consideration, but if not, it should be worthy of a good laugh. These "off-the-wall" concepts will encourage the others to begin to open up and say "what if." Not everyone will stretch their thinking, but if only one person benefits from this creative environment, then you will have achieved one more management success.

Imagination can be a useful management tool and the mechanism to provide the incentive for employees to do better. A manager with a very active imagination can channel this skill to effectively improve individuals and company performance. This performance can generate more business, and in turn, increase financial income for the firm. By simply asking "why not" or "what if," a manager can challenge a team to do better.

Through these brainstorming sessions, a manager can begin to generate a creative environment that can continuously distinguish the firm from the competition. At the same time, the manager is encouraging others to maximize their talents, skills, and knowledge. A manager should strive continually to expand engineering, project management, and technician expertise through personal growth and individual imagination. Just take the time to be creative.

Meeting Ground Rules	Remarks
Start the meeting on time.	Don't penalize those who arrive on time by waiting.
Have an agenda.	This is your road map for conducting the meeting effectively.
Listen!	Don't interrupt.
No side conversations should take place during the meeting.	Wait your turn.
Accept the consensus opinion.	Support the team.
Maintain accurate meeting notes.	Issue to team members promptly after meeting.
Maintain an action agenda.	Agenda should detail deadlines, status, and responsibility for projects.
End the meeting on time.	Don't affect others' schedules.

Figure 4-1. Meeting ground rules.

PUTTING IT DOWN IN WRITING

Another way for managers to encourage success is by expressing their engineering, installation, project management, or service expertise in writing. You will achieve two goals through writing. First, you learn from putting your ideas down on paper. This process makes you think through the idea, concept, and/or project. The details become clearer, and you learn from the experience. I have also found that the words come easily when you are the authority on that particular project. For example, writing about one of your efficient designs or cost-effective installations enhances your knowledge of the project. You will also learn more about the subject matter through your writing. By putting your thoughts and findings down on paper, features of the story become clearer to you regarding why something may or may not have worked and what you would do differently the next time.

71

The second benefit of writing is the personal satisfaction and feelings of achievement you experience. This can be considered a by-product of having an article published. I have never written an engineering article so that people would know who I am. Instead, my writing goal is to highlight a particular project, case study, or concept. With the publication of an article, the customer and/or the company will receive the recognition for a job well done. This management tool is so important to recognize and appreciate. Over the years, I have had the opportunity to be involved in numerous one-of-a-kind hvac/r systems. In return, I have been fortunate enough to put down in writing the results of many of these projects. As a manager, you owe it to your customers, employer, and fellow employees to let others know the value of their investment and the level of effort that went into making the successful hvac/r system.

Writing allows you to receive positive feedback from your peers. In addition, writers can create the myth that they are authorities on a particular subject. However, being able to write about a subject doesn't necessarily elevate you to "expert status," although it does begin to separate you from the rest of the crowd. Sharing your experiences through writing can enhance your standing in the hvac/r community, as well as the standing of your firm. At the same time, you are giving your clients and your company the opportunity to receive national exposure through your writing. When highlighting the results of a successful project, a manager should take the time to maximize the opportunities offered from this project. I think managers, in particular, are obligated to strive for excellence, for the benefit of the company, the employees, and the managers themselves. Keeping a creative firm "under wraps" is not a very good business move. As a manager, you should be responsible for the other employees and the financial success of the firm.

Each and every time a project that began through imagination is completed, get the word out! Take the opportunity to let others know what your team achieved. People often miss the chance to highlight a successful project. If you don't do it, some other company may just take the time to get the word out regarding a similar feat for their own benefit. The by-product of magazine exposure is that you are the authority. More importantly, writing improves your knowledge about a subject, system, or technique. This added knowledge will help you to help others in the company. Employees should also be encouraged to make the time to author an article that documents their hvac/r experience or knowledge.

EXPECT THE MOST FROM EACH EMPLOYEE

An integral part of managing people is a manager's responsibility to encourage each employee to strive to be the best, i.e., the best draftsperson, purchasing agent, technician, pipefitter, etc. You don't have to be a spiritual leader to continuously encourage individuals to "go that extra mile." Sincere support and motivating words of wisdom are all that is needed. In recent years, I have told each and every person that I have been responsible for that I expect the most from them, for their benefit as well as the benefit of the company. For the most part, the workers have been younger than I am and are anxious to advance within the company. I'm not saying that older people aren't anxious to get ahead also; but in my experience, the majority of those employees who are actively seeking advancement have been in their twenties and thirties. It is the period of time in their working careers that they look to gain as much experience and salary advancement as possible. These are their peak earning and learning years. I make a point of letting them know that I understand their concerns, because I have been there myself.

In the past, I have had the opportunity to assist a number of workers with their advancement within the company. With each new person that is assigned to the group, it has been my philosophy to get the most out of them, for their benefit as well as that of the company. In particular, hiring graduate engineers right out of college with no practical experience has been an approach that I prefer to do when possible. By hiring the graduate, you have a person who is "fundamentally trained." This training will be beneficial three to five years later when this person is in a responsible position.

During the first year of employment, engineers-in-training will be a burden on the manager due to their lack of experience with the practical side of hvac/r engineering. In fact, I have found that you sometimes have to intimidate these trainees to get them to realize what they don't know. This problem can be attributed to the fact that these new employees recently invested four or five years and thousands of dollars in college. During this time, they were probably told that they would be a valuable addition to any engineering firm after their graduation! Fortunately, or unfortunately for this new engineer, the hvac/r industry has reduced theoretical engineering into "cookbook" engineering. I can remember one new worker calculating the pipe resistance in a hot water heating system by using Bernoulli's principle. Hours later, when he was completed with the calculations, I showed him a simpler method and suggested he never do that again.

Most individuals will take the initiative to do better and accomplish more if they know their boss has the confidence in them to get the job done. The manager must take the time to point out this work ethic to the worker. By communicating with each employee that your goal is to see them succeed, they will understand why you expect continued performance from them. At the same time, you get them to commit to their goal. If they don't succeed, make a point of emphasizing that it will be their fault — not the company's fault. This is a fair statement based on my experiences in working with other staff personnel. With this statement, I also recall the past success of other individuals, which these new workers can relate to. In addition, it is important they understand that you expect them to fulfill their goals and commitments, since they will be given the opportunity to stretch their talents with work that will challenge them.

As the leader, it is also important to recognize that you can't do all the work yourself or be responsible for all the important decisions. Delegating responsibilities to other employees is essential. If the worker understands that you are counting on them to succeed and that you are working with them, then the task has a very good chance of being accomplished. However, it is still the manager's responsibility to ensure the work is being done, particularly if the employee is being asked to complete a task he or she has not performed before. This means routine checks of the progress being made by the individual. A good way to establish these check points is to meet at the beginning of the project and help establish the milestones necessary to achieve the end product. Using good goal setting procedures, the manager should sit down with the employee and establish project goals. By getting together and writing down the necessary "things to do," a manager can stay involved in the process, while the person responsible for the project is held accountable for meeting the deadline.

Whether the person is new to the group or has been with the firm a number of years, it is the manager's responsibility to expect the most from that person. It is also the manager's responsibility to provide the environment in which each worker can excel. Through the establishment of annual goals, individuals are given the opportunity to succeed. This success should be the achievement of company-oriented goals, and a by-product should be that the worker has grown educationally within the company. Maintaining close surveillance of each worker's progress is essential to the success of the particular project and the advancement of that worker. Using the milestones and dates previously agree upon, the manager will remain in control of each scheduled event, while employees are held accountable for their own performances. In the process, employees are given the opportunity to contribute. A year later, you can point to the person's progress and either agree that the individual has accomplished what was expected of them or the person has not reached the goal.

EXPERIENCE

A manager must be experienced in the complex activities that he or she is responsible for overseeing, whether it is design, installation, maintenance, or service. In the hvac/r industry, managers need to be proficient in the technical aspects of the business if they are to be successful at their jobs. I have often reflected on my education, both on-the-job and academic, to assist me in my management role. I also reflect back to when I was on the other side of the street. It is important for managers to reflect back to what their outlook was when they were fulfilling a particular job description in years past.

Reflection can be beneficial when you are completing an employee's annual review, offering encouragement to an employee, or asking a person to complete a project in a particular time frame.

Experience can be a cost-control tool for a manager when assigning work to individuals. Managers can assign tasks and designate times for completion of those tasks, because, at one time, they had to perform these tasks themselves. If you had the responsibility to complete the assigned task, you should know how long it will take to complete. This is particularly helpful to a new person or a trainee who wouldn't ordinarily be able to determine the time needed to complete the assignment. By stating the goals of the project and what you consider to be an acceptable period of time to complete the job, employees know what is expected of them. Through continued task and time frame management, workers can also form an assessment of their performance, and this information can be useful to both the workers and the manager at their annual performance reviews.

When designating the time to complete a job, the manager must assign a time frame that is achievable by the employee. If you can't do the job in the predetermined period, then you may not be able to expect the worker to succeed with your request. In fact, some people are apt to challenge the manager to perform the job within the scheduled time allotted. This opportunity arose once with a co-worker and myself. I knew the job could be done and I did it, on time. You do not want to create an adversarial atmosphere with your time frames. Instead, the purpose is to assist the individual with what is an acceptable level of effort and to encourage them to strive towards doing a good job faster.

An experienced manager should share his or her professional growth experience with employees, so the staff can appreciate the leader's background. This past experience will often recur time and time again, because you are not the first person to work your way up through the ranks of business. Almost weekly there will be an occurrence that a manager finds similar to an incident that happened to them years earlier. I believe reflection can help you lead the staff forward, and the by-product of sharing your work experience with others is that they can learn from you. There are very few managers who consistently take the time to share their practical experiences in the business. This sharing of your business past is another form of encouraging workers to be the best they can be.

You should continually show interest in the employees, their goals, commitments, and, sometimes, their personal problems. You should also be comfortable with the company role and yet not be aloof in the position. Sharing your experience is essential if you expect to get the most out of employees and, in particular, the trainees. Having gone to college evenings, I recognized the benefits and rewards of knowing the fundamentals. What college didn't teach me was that as an employee and as a trainee, there is a practical side to hvac/r work. As the manager, you should bring that practical, individual education to the work environment. The fact that you've walked in their shoes brings a sense of reality to the theoretical teaching of hvac/r.

It can also be fun to prove you are as good as you've implied when you designate the project time schedule. Rolling up your sleeves and getting the work done can be a refreshing change from managing people. Another example of this "can do" attitude is the story of an engineer, with whom I have had the pleasure of working. He started in the business as a refrigeration mechanic. He went to college at night and is a registered professional engineer today. When we had a problem with a refrigeration system, he was sent to the site to meet with a service technician and troubleshoot the problem. Because of his background, he brought his own refrigeration gauges to the site to test the system. This was probably the first time this service technician ever met an engineer at a jobsite who brought his own gauges!

SUMMARY

The first step towards maximizing your leadership is to **get the job done on time and under budget!** Let your actions speak for you. This may sound like easy advise, but it takes years to establish yourself as a person who can get the job done. People have to get up every day and do the best they can at what they are paid to do. Leaders aren't created overnight. The same can be said for effective managers. A manager must take charge each and every day. By setting the example, this person can establish himself or herself as the standard of performance each worker can strive to match. Along the way, a manager needs to inject a sense of humor and a personal touch when working with other employees. Life is too short to be constantly under stress or to place workers in a continuously stressful environment. Employees want and need to work in an office environment where people enjoy each other and appreciate each other's skills. A good leader can make this happen.

Another issue that a manager must control is frustration. I find that, as the manager, you are expecting the most from each and every individual for which you are responsible. Continually stress that you expect the most from each employee, for their benefit as well as the company's benefit. Even with this philosophy, there will be the occasional person who doesn't do the best that he or she can. It's times like these that disappoint me, as well as those times when the workload becomes excessive and frustration sets in. As the person in charge, it is acceptable to share your disappointment with these individuals, because you know they can do better. Managing the workload so that "overload" conditions can be kept to a minimum will also aid in keeping frustrations to a minimum. Don't let overload be an everyday way of doing business. Imagination can be a valuable tool for you to work with when balancing work, deadlines, and staff shortages. Seek out engineers, CAD operators, technicians, etc., from outside the office and establish a network of subcontractors who are able to respond when workload peaks are in your forecast.

Imagination is also a valuable tool when performing work in any of the many areas of hvac/r technology. Through team "think" and brainstorming sessions, a company can grow, achieve increased profits, and establish itself as unique and exceptional. Just as important is putting down in writing the positive experiences of a project. Capturing the story, concept, and team effort is a responsibility of the manager. The leader should take the time to educate the industry regarding innovative technology or creative problem-solving. By doing this, the manager is letting others know exactly why customers have wisely invested in their facility operation, expansion, and/or renovation.

Being active in your manager's responsibilities, effectively working through the daily stress of the business and its people, and doing it with a sense of humor will certainly assist you in maximizing your leadership. Leading creatively and sharing your experiences with your customers and workers is an excellent way to demonstrate a job well done. Continually renewing the excitement for the job, the company, and the product produced is necessary every workday if a manager is to continue to pilot the team. You can't direct from the rear. If you do, it won't be long before someone else will want to step to the front and take charge, just like you once did.

NOTES

[1]   Blanchard, Kenneth, Ph.D. and Spencer Johnson, M.D., *The One Minute Manager*, William Morrow and Co., Inc., New York, 1982.

[2]   Ibid

Chapter 5
The Tricks of the Trade

Every business has its "tricks of the trade," and managing in the hvac/r industry is no different. If you are just beginning to manage people, then it might be helpful to purchase a small notebook and jot down the techniques or "tricks" that work for you. Consider indexing the notebook into separate categories that are appropriate for your business, such as rules of thumb, personality traits, non-think work, budget values, and safety tips. Don't try to keep everything in your head. In the future, you may not need these reminders, but until then, keep your experiences in a notebook.

You may change these categories or add to them as you increase your knowledge of the business. Note taking can help you track an idea you are interested in developing into a rule of thumb. By continually jotting down the data, a pattern may develop that you can formulate into your own trick of the trade. This type of note taking allows you to compile historical reference data, which you will need throughout your hvac/r career. For example, I have found that mechanical rooms usually require 7% of a building's floor space; constant volume hvac/r systems require 1-1/2 cfm (cubic feet per minute) per square foot, while variable volume air systems require 1-1/4 cfm per square foot; pump installations cost $5000 per pump for an end suction pump installation, while split case installations cost $10,000 per pump; and you should always look for an eye wash station at a chemical treatment testing table. Obviously these examples touch on a broad spectrum of issues, but an hvac/r manager will need to develop these "quick check" values depending on whether he or she is responsible for engineering, construction, service, maintenance, etc. Never underestimate the value of experience and the importance of using tricks of the trade to quickly assess any situation.

The paragraphs that follow contain some tricks of the trade that are in no specific order of importance. Instead, they are a random selection of keynote issues, useful for all managers in the hvac/r industry.

CONSTRUCTIVE CRITICISM

Managers are often called upon to offer helpful advice to individuals, sometimes on a daily basis. When initiating a discussion regarding a work issue, always remember that **timing is everything**! More often than not, the individual isn't looking for your advice or opinions. As the manager, you constantly need to monitor the progress of each and every project within your group. Most employees respond to constructive criticism in different ways (referring back to your notes regarding personality traits should be helpful in these situations). Although most workers are willing to listen and heed your advice and experience, there will be others who will scoff at your input. For those particular individuals, a manager needs to be more creative in order to be successful. Talking to the person in the privacy of their workspace or possibly in a neutral space such as a conference room can provide the needed environment for constructive criticism. Don't request that the person come into your office. This may be intimidating, particularly to an individual with a negative attitude.

A manager must be **pro-active** when it comes to constructive criticism, like a cheerleader giving encouragement to the team. You need to routinely speak to the employees with the positive direction and advice necessary to keep the staff "up" for the work. Positive constructive criticism is certainly easier to address with a person than negative constructive criticism, but the ability to discuss those things you are unhappy about is just as important. As a manager, you should say what you believe and not what you think! Sometimes a manager will try to anticipate the reaction of the individual receiving the constructive criticism and try to think of a method that may be more palatable. If you think too hard about how this person is going to react, you may miss the opportunity to clearly present the issue at hand. State the facts to this individual as politely as possible during a one-on-one conversation. Emphasize the issue as you see it, and state how you believe it should be handled. Listen to the person's response, but make it clear exactly what you believe to be the correct way of dealing with the problem. I'm not saying that the employee is always wrong and the manager is always right. What I am saying is that as a manager, your message for this person is to do better, and you are simply offering methods to achieve this goal.

Don't ever censure a person in front of others. You will embarrass not only the person you are talking to, but also the people around you. Under this condition, a person will most likely go on the defensive, and this will only lead to a confrontation. Also consider the individual and how he or she is feeling and acting. I have observed some workers who are usually not in a good mood early in the morning. This type of person will often be very defensive at the beginning of the workday. Waiting until after lunch may be a better time to approach the individual. Later in the day, you will find them in a better frame of mind. Don't make the situation any more difficult than it has to be! Waiting a brief period of time won't prevent you from addressing the issues and achieving your goal.

Another "don't" for managers to remember is never to lose your temper when giving constructive criticism. Getting into an argument will never achieve the focus of the meeting. Be prepared when you meet with an employee, and concentrate on the task at hand. Losing focus in the discussion and attacking the individual's character will accomplish nothing, and in the long run, the person will lose respect for your leadership. If the person chooses to attack your character or abilities, let it pass. Never respond to those types of statements. As the manager, it is your job to lead, direct, and guide. It is also your job to get the most out of employees, for their benefit as well as the company's benefit! Constructive criticism goes with the job.

THE BAD LUCK SYNDROME

According to the dictionary, a syndrome is "an aggregate or set of concurrent symptoms indicating the presence of a disease." I once worked with a person who had acquired a negative attitude that proved to be infectious to those who worked with him. This man, whom we will call Bill, was talented, educated, and experienced and had a lot of pride in his work. Unfortunately, Bill assembled a collection of pessimistic traits that would form what I call the **bad luck syndrome**. At first it wasn't obvious to me that this disease existed. I brought this to his attention on a few occasions and made subtle comments about his job outlook, hoping that he would overcome this negative attitude. After a few months, it became apparent to me take he was inherently pessimistic and that his perspective was spreading to the other members of his design team. Each week I would have a scheduled review of the various projects in the design phase and the construction phase, and each week Bill would complain about the progress, or lack of progress, being made on his job. It seemed that the architects were against him, the staff wasn't working hard enough, etc., and he just wasn't satisfied with the design. At these same meetings, other project engineers also voiced their problems, but they usually were very positive about the end product results. I also observed that the members of Bill's team were equally as pessimistic. At the same time, I noted that the other teams appeared to enjoy the projects they were assigned to. Not only was their outlook positive, they also accepted the challenges of the job. Bill and his team complained while they worked. They seemed to have a lot of "bad luck" with their projects!

From my observations, I noted that if the leader is not optimistic about the job and doesn't encourage the team to move forward, then the team isn't going to move forward on its own. In Bill's case, the team followed him through the completion of the job, complaining all the way. Whether you are a manager or a project engineer, you must be out front leading! The person in charge needs to create excitement about the project, each and every work-day. As a manager, you cannot complain about a job constantly and expect your team to make up for your negative attitude by passing you by. They know you are the person responsible for bringing the project in on time and within budget. They are not in the position to bypass you and move forward with excitement to meet the team goals. It is all right to show frustration with a problem, but it is your responsibility to overcome these obstacles and continue forward. In fact, a little humor can do a lot for a project that is bogged down with project conflicts. Sometimes it is good to stop and assess the situation and laugh at what you did wrong.

Since I met Bill, I came across an informative and enlightening audiocassette tape that examines this negative personality type. After listening to this tape, I requested each of the people in my group to do the same. I call the cassette the "bad luck tape." The narrator of the tape starts out by discussing his past experiences working with prisoners in a penitentiary. It appeared to the inmates that many of them were in jail due to "bad luck." The narrator goes on to discuss the reasons why these inmates believe their "bad luck" got them thrown into jail. It seems that these individuals truly believed that they shouldn't be in prison and that it was someone else's fault for their misfortune.

How often have you heard someone tell you something didn't fit, didn't work, or was someone else's responsibility? When something doesn't go the way they'd like it to go, how often do these people assign the blame before they offer a solution? These are examples of the bad luck syndrome and a negative attitude. If there is a problem, the solution should be the first issue on the agenda. But many people will attempt to clear their name from the problem before they address its solution. Having had each person in my group listen to the bad luck tape, I expect the solution first when a problem arises. When the first response is to clarify who is inno-cent and who made the error, I always make a point of "sympathiz-ing" by saying "that's bad luck!" The employee then knows from my comment that I'm not interested in who made the error, but I am concerned with the solution. I eventually want to know that the person who made the error is aware of how the problem developed and how it was solved. It is important that people learn from their mistakes and the mistakes of others. Bill's team was always con-cerned with not making mistakes; but when they did occur, the team was quick to point out that they were not responsible for making the error. This is the wrong attitude for a lead engineer or manager to project to the team members.

NEVER BACK SOMEONE INTO A CORNER

The hvac/r business is a surprisingly small community of workers, and within this community, you will frequently cross paths with the same people. Often, you will hear of individuals who are very good at what they do or who may not be as good as they would like others to believe. This is the same for your co-workers, some of whom may be very talented or very limited in their technical skills. No matter where these people are, when working with them, you must always be courteous and professional. It serves no useful purpose to be vindictive towards another employee, a business competitor, or a customer's employee.

A useful phrase to remember when dealing with someone who doesn't share your opinion or who is simply incorrect is to **never back them into a corner they can't get out of**! This trick of the trade can be applied when working to resolve problems with all types of people. Until the problem is resolved, don't confront the individual with a solution that can be interpreted to mean that he or she was responsible for the error. No one intentionally makes a mistake. It is not only "bad luck" to identify the culprit, it is also not very professional. The problem has to be resolved first, because it is paramount to moving forward. Challenging the individual will only hurt your relationship with that person and possibly the relationship of the others at the meeting. Once the solution has been determined, most people can recognize who made the mistake. You don't have to confirm the guilty party. In addition, no one is perfect. You can bet that some day you will be on the receiving end of a problem, and the person giving out the constructive criticism will be the person you had "cornered" in the past.

One way of avoiding this adversarial dilemma is to meet and discuss the issue separately with the other person prior to any formal meeting that will have others in attendance. Together, you both can resolve the problem. If the other person is uncomfortable with the solution because of how it may reflect on him, help him through his problem. The sole purpose for meeting with the individual is to resolve a problem, and that problem isn't who is right and who is wrong. Try to be as diplomatic as possible so the agenda doesn't turn into a liability issue. Neither of you will benefit from the meeting if this occurs.

Confrontations, such as two companies addressing the same issue, are quite common when a customer brings in another consultant, contractor, or service company. This can occur because the customer doesn't have the in-house expertise or because the problem has created a credibility issue with your firm. In either case, you may have to meet and work with this other firm. Unfortunately, this other firm may not be as sensitive and diplomatic as your company would like. When an adversarial relationship develops, it is important that you don't stoop to their level. Some people see this as an opportunity to demonstrate their expertise, knowledge, and years of experience. In addition, they may even try to back you into a corner, so they can be the hero of the meeting. This may be the opportunity your competitor has been waiting for, a chance to make themselves look good at the expense of your company.

When this situation occurs, and you can be assured that it will occur at some time in your management career, listen to the person's opinions. Stay focused on the facts, issues, and findings of the other individual. Stay on course to resolve the problem. Don't take the accusations personally, even if they are directed at you. If the other person is correct, then the problem is solved. However, the solution is not always that simple. After all, if the competition can solve the problem, then your firm should have been able to come up with the same answer prior to them coming on board. What usually occurs with the other company is that they don't receive the full scope of work, and it is not made clear to this second company. The second company may not have been given all the facts, and/or this second firm erred with its own assessment. No matter what the reasons are that make the second firm wrong, you should not converge on their mistake in the same manner that they attacked you. Instead, stick to the game plan of being professional. Make a point of complimenting them on their report, even if it was insufficient and incapable of truly resolving the problem. Let them "off the hook" graciously and professionally.

The day will come when the other person is on the "hot seat," and your firm is brought in to give a second opinion. In this situation, it does not serve the customer for you to continue your feud with a competitive firm. It is also not of any benefit to turn the client's meeting into a fight of two people still arguing over an old problem. This is a new problem in need of a resolution. This meeting is an opportunity for your firm to serve your competitor's customer. Resolving the dilemma, even if it is only to confirm the other person's position, can be a victory for you depending on how you present yourself and the firm. Your goal should be to solve their problem first, and then leave with the customer wanting to work with you in the future because you are proficient at what you do.

ROUND TABLE DISCUSSIONS

An invaluable resource and training tool, seldom used in the hvac/r industry, is the **round table session**. In today's competitive arena, engineering firms don't share technology with other engineering firms; construction firms don't share production information with other construction firms; and service firms don't share information with other service firms in a formal round table setting. Individuals from different firms may meet occasionally, but seldom do they share their knowledge on a regular basis. A regular basis may be semi-annually or annually, and it should be a very informative meeting of a select few.

Managers should have the confidence in their abilities and the abilities of their firms to benefit from a meeting of a select few from other companies. The emphasis on the "select few" is to develop a core of members at the beginning. The process needs to be managed so the meetings are beneficial and so the process will continue for years to come. This round table gathering should be above and beyond the monthly meetings held by professional associations. The forum can be a company and its customers, rather than a group of similar hvac/r companies. Getting your customers to participate in a regularly scheduled round table session provides you with invaluable information and resources from "end-users." In addition, your customers will be able to meet each other and learn from their professional counterparts. At the same time, compliments directed toward your firm are made by customers in the presence of other customers.

To begin establishing a roundtable session, focus on your repeat customers. Professionally, this is a great place to start, because you are assembling a group of people to whom you provide a service. I have always believed that an engineer with a creative mind and engineering skills is still very dependent on the operations personnel for the success of the project. No engineer can truly claim the complete success of a system simply because he or she designed it. It's people like the director of engineering who make the system work! Bringing these key people to a meeting to discuss the agenda can be invaluable to a design firm. The same can be said for a construction company or service company that invites the same facility personnel. As the manager, you should take a pro-active approach to the concept of a round table session. Initiating and maintaining this kind of forum will increase your knowledge of hvac/r technology by allowing you to hear other perspectives.

Invite the participants to this round table session by telephone, not by letter. Personally request their participation, and note the benefits that you will receive from their input and what they will receive from meeting with others and sharing their facility strategies. Identify the other participants once they have pledged their participation. Issue a draft agenda, and request their immediate response for finalizing the forum. Ideally, it would be preferable to formulate a **mission statement** at the first meeting. From then on, the team would focus on this as the ultimate agenda. In Chapter 1, we discussed the tools and techniques for conducting successful meetings, such as establishing meeting ground rules, setting an action agenda, and setting goals. All of these procedures will provide you with the tools to manage and record the meeting activities. The frequency of these meetings also needs to be agreed upon. The best that can be expected will probably be a semi-annual meeting due to scheduling conflicts.

A by-product of this assembly is that you can establish yourself as a source directly with your customers. This gives you the opportunity to separate yourself from all other managers who go to work every day and consume themselves in their daily activities, without thinking of how they can improve themselves through customer relationships. When meeting with customers, ask them the following questions:

- How can we do better the next time?
- What worked well, and what didn't work well?
- Was the design too theoretical when compared with the "real world" application?
- How can the construction cost fit into the operating budget better?
- What should be specified and carried in the annual maintenance and operating budget?

From this education, you can better serve your customers while sharing this information with your staff. It is imperative that when you start this process, you are able to continue it on a regular basis. You should make sure the information is shared, the meeting notes are issued, the agenda updated, and the participants identified. Approach this forum like any other regularly scheduled conference or jobsite meeting. Address it professionally and proficiently. Don't waste your customers' time!

REVIEWING DEPARTMENTAL WORK

A manager needs to be continually involved with the activities within the group. This group can be a service, construction, estimating, engineering, or maintenance department. No matter what group you are responsible for, maintaining control is crucial to the staff's success. Whether you are a project manager who is managing a staff of people or a manager who is training a new project engineer, consider the 8:00 a.m. and 1:00 p.m. review process. Years ago I was taught that you should check with each person at the start of the day and right after lunch. The idea behind this four-hour spot check was that an employee would not be able to "go off course" by more than four hours. This is particularly useful when working with a trainee. The person who taught me this management tool was a strict disciplinarian, and a by-product of this strategy was that it got the staff working at the start of the day and immediately after lunch. He would come up to you and ask you where you were with the project and assess whether you were "on course." If you were talking to someone when he approached, the other person would feel obligated to go back to his or her workspace. When he walked away, there was no one with you to talk to, so you usually began to work. The same thing would happen immediately after lunch. There was a method to his madness!

The next step in this process of reviewing departmental work is to schedule a weekly meeting with the other personnel in management positions. It is important to keep this group to a minimum, even though egos may be hurt if everyone is not invited to participate. The weekly agenda should be established at the beginning of the meeting and should be updated immediately after the session ends. The standardized agenda should include the date to distinguish it from the older copies of the voided agenda sheets and to ensure that everyone is working with the same data. The agenda should also include the complete list of active projects, job numbers, and the names of those responsible for the projects. Because most jobs have multiple deadlines, this information may be cumbersome to compile on one sheet, unless the information is spread across a horizontal sheet of 8-1/2" x 11" paper. The results of the session will serve as the action agenda for the current week. The decisions stated and/or agreed to at the meeting will be the road map of activities for that week. These same project milestones and deadlines will have an impact on the weekly manpower and could affect the weeks to come if schedules are not met.

The weekly meeting agenda should be confined to one page, if at all possible. Two pages are the most that you should work with. Too many pages may be an indication of an excessive agenda, resulting in a very lengthy meeting. As the manager, you should be interested in the milestones and benchmarks. You should not scrutinize the finer details of any one project, because that is the responsibility of the project manager, foreman, project engineer, or supervisor. Most importantly, you want to leave the meeting believing that the other members have a positive outlook and are in control of their projects. I was on_e told that any more than two pages means you are not getting the work done!

A **monthly manager's report**, Figure 5-1, is also an important document to maintain. This type of report is an appropriate way of keeping your boss informed. The report should address the month's activities and expenditures, as well as those for the year. It should also include the manpower available and required, the departmental goals and status, and any current events.

Beginning with the activities, you have your weekly agenda for project activities and associated financial data. This should be compared to the business plan, which should state the volume of work and gross profits for the year. Departmental goals are the company-related objectives that you expect to achieve by year's end. Current events are the in-house activities you consider to be news worthy. An employee who passed his or her CFC certification, a worker returning from maternity leave, etc., are a few of the general interest issues that should be shared with other departments. The month's activities, project cost control, manpower, and department goals are also an integral part of measuring your performance as a manager. The business plan and your management strategy need to be in parallel with each other. Your success and the success of the company are dependent upon everyone pulling in the same direction! If you are a first-time manager, you should schedule a meeting with your boss to develop and agree upon a reporting system that will be informative and a means to measure your efforts.

STANDARDIZED WORK FORMS

The **quality process** has probably done the most to bring standardization to the forefront of hvac/r management. This simple tool has made my working career so much easier, beginning with my first **drawings checklist**, which was developed during my first year in the business, Figure 5-2. I learned early on that it was very helpful to develop a checklist to document all the tasks I had to consider when drafting and designing an hvac/r system. The idea is to list all the possible "things to do" regarding the drawing on one sheet rather than keeping all these drafting/design duties in your head. As the work is completed, I shade in the block. If the work is only half completed then the block is only shaded halfway across.

Operation & Maintenance Manager's Report

Date: May 5, 1994   Prepared By: H. McKew

Subject: April, 1994 and Year-To-Date
         O & M Department
Distribution: President
         Vice President

FINANCIAL:

April		Year-To-Date		
Budget	Actual	Budget	Actual	Comments
$5,000	$4,725	$19,000	$17,300	Includes O.T.

STAFF REQUIREMENTS:

April		Year-To-Date		
Budget	Actual	Budget	Actual	Comments
4	4	4	4	No Comments

PROJECT STATUS:

Job #	Job Name	Remarks
# 9401	Project X	In Project Close-Out Phase
# 9402	Project Y	Safety Approvals Required
# 9403	Project Z	Two Weeks Ahead of Schedule
# 9404	CFC Issues	Certifications This Week
# 9405	IAQ Report	TAB Report Overdue

DEPARTMENTAL GOALS:

Goal #	Theme	Remarks
(1) Develop a Master Specification	Computerize Process	Two Weeks Ahead
(2) Update Standard Work Orders	Computerize Process	One Month Ahead
(3) Issue Energy Graphs Monthly	Computerize Utility Bills	Two Weeks Behind
(4) Standardize Emergency Plans	Input On Word Processor	On Schedule

CURRENT EVENTS:

Sylvie had her baby and will be returning to work the week of June 2nd. Jim has passed the E.I.T. examination. Bob will be on vacation next week. Please coordinate Goal #2 with him.

Figure 5-1. Example of a monthly manager's report.

A quick look and you know that the task is only half done. If the item isn't applicable, I shade in the whole block. This indicates that I considered the work but it didn't apply. The by-product of this checklist is that you can quickly identify the **non-think work** and have those chores ready to be completed whenever someone in-house is looking for drafting work.

Drawings Checklist							
Item	**___% Complete (Shade box)**						**Remarks**
Drawing title							
Title block							
Room names and numbers							
Scales on drawings							
Key plan							
Equipment requisition							
Legend							
General notes							
Engineer's stamp — signed							
Unit location							
Unit weight (issued) with locations							
Electric data (issued) with locations							
Floor drains							
City water and backflow preventers							
Ductwork distribution							
Duct sizing							
Registers and diffusers sized with direction blow							

Fire dampers								
Smoke dampers/ detector								
Plume dampers								
Access panels and doors								
Sound lining (attenuation)								
Louvered doors								
Cfm and gpm								
1/4" duct shafts and details								
Mechanical room ventilation								
Corridor make-up air								
Piping distribution								
System schematics (air/pipe)								
Trap sizes								
Pipe sizes								
Steam to owner equipment								
Pipe expansion loops								
Anchors and guides								
Thermostats (humidistats)								
Humidifiers								
Flow arrows								
Drain valves								
Air vents								
Drain pipes								
Flow meters								

Coordinate with electric and plumbing								
Service access/coil removal								

Figure 5-2. Example of a drawings checklist.

The quality process supports this concept, because it is repetitive. If you develop a standard form with which to manage, engineer, start systems, or maintain equipment, then this sheet will be your first step towards quality control. Anything that is repetitive is measurable! In order to have quality control you need to be able to measure the results. Through measurement, a manager can analytically determine that the task is being performed each and every time. Using a checklist, you can quickly log in the duties as complete or incomplete. Standardizing the process is an integral part of good management skills.

This is particularly important when reviewing another's work. For example, when there are a number of engineers in the group, it is very important that they document the design using the same standards. This isn't to say they can't be creative with their hvac/r engineering. All it means is they should adhere to the same standards when documenting the information, i.e., they should use the same requisition sheets, electrical data sheets, drawing sheet size, etc. As a manager, you cannot afford the added time it would take to review a project that is documented on a format you are not familiar with. Allowing people to use their own guidelines, without your approval, will only slow down the review process and leave open the opportunity for something to be missed. The more people you are responsible for, the more you need standardization.

In addition to checklists, the use of a **matrix** can be an invaluable standardized tool, Figure 5-3. When evaluating hvac/r systems, maintenance options, etc., this simple horizontal/vertical analytical process allows you to compile the data onto one sheet. By listing the hvac/r systems horizontally across the top of the matrix and listing subjective/objective characteristics and historical data verti-

cally down the left side of the matrix, you can evaluate your options. This is a process a manager can implement into the group and expect the staff to use. The concept is easy to implement and flexible enough for different groups, such as service, construction, and estimating to use. Also, understanding the results is quick and easy.

The following is an example of the objective characteristics used to rate various hvac/r systems:

- Initial cost (must be less than $8.00/sq ft)
- Stationary engineer (preferred)
- Equipment out of the workplace (preferred)
- Temperature 73°F ±2°F (required)
- Humidity control (not required)
- Sound level at 35 to 40 NC (required)
- Air filtration at 50% efficient (preferred)
- No equipment on roof (required)

The following is an example of the historical data used to rate various hvac/r systems:

- First cost
- Maintenance cost
- Operating cost
- Life cycle cost
- Expected down time
- Energy consumption

The following is an example of the subjective characteristics used to rate various hvac/r systems:

- Appearance
- Maintainability
- System reliability
- Flexibility
- Comfort
- Occupancy/move-in date
- CFC/HCFC
- Available floor space
- Engineered smoke control

By listing these characteristics and data within the matrix, an engineer can then give a numerical value to each of the hvac/r systems. The values assigned to the characteristics and data can then be added. The hvac/r system showing the highest total points is the most desirable.

A matrix can be a manager's best friend when it comes to analyzing a complex issue. The standardized process allows you to easily follow the methodology used by the person responsible for the study. The summary is compiled into a brief, one page statement that allows you to agree or disagree with the results. It has been my experience that customers find this concept easy to read and understand. They don't have to read through an in-depth report to get to the bottom line. The same can be said for a manager who has to review this same report.

Objectives/ Characteristics/Data	System 1	System 2	System 3	System 4
Initial cost (less than $8.00/ft^2	4	2	2	3
Equipment out of workplace	4	3	1	3
Air filtration at 50%	2	3	3	2
Sound level at 35 to 40 NC	3	3	2	4
Maintenance cost	2	3	4	3
Energy consumption	2	2	4	4
Appearance	3	3	2	3
Flexibility	3	3	2	2
CFC/HCFC	3	4	4	4
Comfort	2	3	3	4
System reliability	3	3	3	4
Occupancy/move-in date	3	3	3	2
Totals	34	35	33	38

Rating Scale: 4 = Excellent; 3 = Good; 2 = Fair; 1 = Poor

Figure 5-3. Example of a standardized matrix.

CUSTOMER SERVICE

There are a few tricks of the trade that can serve you well when continually serving your customers. The first tip is to **call the customer on the coldest and hottest days of the year**. If the hvac/r system is operating satisfactorily, there is no doubt your design or installation works. It should be standard operating procedure to telephone the customer on a **design day**. The client will probably appreciate the concern and interest, and a by-product is that the client has been made aware that the system is operating per design. At a later date, when the system is not operating properly, the customer knows that it had been operating satisfactorily in the past. This doesn't mean you don't help the person on the other end of the line, but it does eliminate the concern as to whether the system has sufficient hvac/r capacity. The most difficult problem with this concept is getting the project manager or lead engineer to make the first call. Some people believe if they don't hear any complaints, then everything is fine. Customer service has proven this myth wrong, because in today's marketplace, you can't afford not to stay in communication with your customers. If you don't, there are others who will gladly make the call to find out whether your design is operating satisfactorily!

A second tip related to customer service is to make sure the facility engineer, director of engineering, and the operating and maintenance staff get their due share of recognition. No matter how innovative your engineering, installation, or servicing has been, the building operators are the people who have to keep these hvac/r systems going. Often, you or people in your group will overlook the work of these other individuals. But it is the facility staff who has to make the system operate with minimum problems and complaints. They can make your job very difficult if they don't think you are good enough for the task. When writing an article for publication, pursuing an engineering award, or considering a creative idea, I always look to the building operators for their experience. In return, it is imperative that they get their due share of the credit.

A suggested writing tip for when you or a person in your group has to correspond with a client is to review the text when it is complete and ask the question "so what?" With each paragraph of the letter, the reader should be able to clearly understand the message you are sending. What is it you are trying to say? Very often, when a person is writing a letter, memo, or report, the writer usually knows exactly what his or her message is to the customer. The dilemma arises when you try to put your thoughts into words. Your subconscious already has the answer as you write. As a result, you may be unclear about your meaning. Try having someone else read your letter before you send it out and see how your writing is received. Don't try to justify your writing! Appreciate the feedback from the other individual, and recognize that if the text isn't clear to this person, then the individual receiving the same text may have the same questions. Continue this process with each letter or report you write until the reviewer stops questioning the content. After that, if you don't want someone to read each and every letter, spot-check your writing with a random request for someone to review your writing.

When I first started writing letters to customers, the chief engineer was responsible for reading each letter I wrote. From his review, he would ask questions, and if the answers didn't jump off the page, then it was apparent that the message was unclear. As a manager, you should also ask "so what" when reviewing the letters and reports that are issued from your department. If you don't understand what the person is asking, stating, or summarizing, then the customer may have the same query.

INTEGRITY

Probably the most important tip you can pass on to other employees is to **maintain your integrity**! When working with a client, always preserve an honesty towards your work and work relationship with the client. No matter what the product is that you provide to your customers, it must be delivered based on your best effort to perform the work. If you can't do the best job possible, get someone else to do the work. Don't compromise your commitment to your customers. They have entrusted their needs in your hands. Anyone can do an average job, so you should strive to do the very best.

Stress to your clients your commitment to quality performance, and back up your words with results. Managers who distinguish themselves through their actions inherently establish themselves as individuals who have character and who can be counted on when a customer has a need. As a manager, this same message needs to be conveyed to each and every worker in the group. Always remember that actions speak louder than words! Each service product within the hvac/r industry is produced by a small community of people, and employees should recognize that word travels fast within this arena. A lackluster performance by workers in-house or with a customer will stay with you for years to come.

The same can be said regarding your relationship with your fellow employees. Always be honest and sincere. Whether someone agrees with you or not, don't lie or exaggerate your point of view. Some say there is a fine line between promising and committing to an issue. Words can come easy to some. Don't promise something you cannot complete. As a manager, this is paramount to your success. The person in charge must possess credibility that others will respect. No one can lead effectively if the followers don't believe in that individual. A person once said to me "I don't agree with you, but I know you're not lying to me." That was a subtle compliment to my position regarding a specific topic. You can't expect people to agree with you all the time. The best you can achieve is that they respect your point of view, even if they don't concur with your position on the matter.

Make integrity a daily occurrence. Speak up for what you think is correct. This includes offering constructive criticism to the staff of people you are managing. When encouraging someone to perform at a higher level, emphasize your genuine interest in their success. Some managers may have a problem with encouraging a person too much, because the person may then want more responsibility, money, or a higher position. Some managers may even be concerned about the other person taking their job if the person receives too much recognition. As the manager, it is your responsibility to get the most out of each person. Setting a limit on the worker's performance serves no useful purpose. If someone is good enough to take your job, then that should happen. Chances are, if you can get someone to perform at your level, then you are ready to move up yourself, and you have developed your replacement in the process.

Never let outside influences affect how you manage. The one thing you can control is how honest you are regarding your job function. In the process, good managers will inherently establish job security through their ability to manage, not through how sincere they are. Integrity is an integral part of being a manager. No one will continue to be successful at their job if they do not maintain principles of decency. Disturbingly, there is a minority of people in responsible charge whose behavior contradicts their management directives. For them, there is a double standard. Hard as it may seem, these managers usually have lost sight of the needs, opinions, and values of others. They consider themselves to be the select few who understand what is necessary to succeed, and they lose sight of the team effort. Routinely, these leaders will project a "do as I say and not as I do" message. Somewhere along the way, their corporate vision became clouded with success. After that, the influence of power eroded their integrity and standards. Success can do that to some people. Don't ever think you are above the rest because you are the boss.

SUMMARY

Experienced managers, particularly a person with a creative mind, will continually develop many tricks of the trade. These handy tools will help you to be a better manager and leader. Always strive to do your job better. Quite often, people in charge will overlook management details that could make them better managers, as well as better people. Compiling historical reference data is a great beginning. As an hvac/r system designer or engineer, check figures and budget values will prove invaluable to you. If your specialty is service work, similar "tricks" will help you troubleshoot a system, override an automatic temperature controller, or estimate the volume of glycol solution needed in a water system for freeze protection.

Timing is a useful "sixth sense" for a manager to develop as part of his or her management skills. By being aware of your timing, you can improve your ability to communicate with your employees. This same awareness applies to regulating the negative attitudes that workers within the group may develop. Don't accept bad luck stories as acceptable reasons for mediocre performances. Separate the problem from the person making the mistake, identify the error, and correct it. After the problem is corrected, work with the employee to assist in improving his or her performance.

Maximize you firm's exposure by utilizing the concept of round table sessions. Round table sessions are essential to maintaining good client relations and will provide you and your company with unlimited, useful information. The round table approach is a great learning process. This gathering of a select few customers allows the firm to showcase what it has to offer and its professionalism. Managers need to be pro-active when establishing these tricks of the trade. The by-product of gathering a group of experienced clients together is that you will be able to understand the hvac/r industry from their points of view.

Reviewing the work within your department requires a creative mind. Standardization can be the cornerstone to this process. Beginning with the weekly, monthly, and annual reports, standard agenda items will ensure that the goals are met. Consistent reviewing procedures and repetitive checking of work can accelerate your capacity to manage. Maintaining a "things to do" list, using a matrix table, and compiling non-think work will complement the review process.

Recognizing the knowledge and responsibility of the facility engineer is another useful tip managers should pass on to the other members on staff. Communicating in writing with the facility engineer and others is a daily activity. Writing in an effective manner will often require that someone review your letters and question what you mean if it is unclear. Getting a second opinion can improve your writing skills, as well as the skills of others in the group.

Beginning with an appreciation for a client's point of view, a simple telephone call on a design day can be an invaluable customer service trick of the trade. Integrity is directly linked with this appreciation and awareness for clients and co-workers, and it is a priceless trait all great leaders should possess. Along with integrity, you should also work to continually be the best in the business through consistent and exceptional hvac/r skills. All these tricks of the trade are the building blocks to being a better manager.

Chapter 6
People

A manager must be sensitive to the heartbeat of the group and the company. It is important to appreciate and understand the different types of employees and personalities that make up the workforce. These people may be grouped together at a project site, distributed over miles of service territory, or within a single office environment. Failure to recognize the individual inevitably will result in a drop in employee performance. Sensitivity to the workers, while looking out for the company, is an integral part of managing people in the hvac/r industry. You need to have a sixth sense towards the staff of workers for which you are responsible. Feeling that business heartbeat is an important part of your job. Knowing when to reprimand, compliment, and encourage are just a few of the people skills managers need to possess if they are to truly be successful.

A manager should listen when workers talk. Don't interrupt. Let them have their say, and concentrate on what they are saying, not what your response will be. Caring is also a part of managing these individuals. During your daily routine, employees need to be able to come to you and express their problems, concerns, and needs. Staff members may be experiencing personal problems, such as divorce, physical abuse, and/or drinking problems, and as a manager you will be confronted with these and other problems. You need to be prepared, not necessarily to solve their problems, but to give assistance, direction, and/or support. It is very important that they be able to come to you with their problems, so you can keep the work on schedule while they work out their troubles. From a company's standpoint, the job still needs to be done.

It is also important that your employees feel comfortable working with you and are able to talk to you. At the same time, you don't want to get into a position where a worker is continually burdening you with his or her problems. Show an interest, but be able to determine if their predicament is going to have an impact on their performance at work. If it doesn't impact the office, deflect the issue back to them and get back to work.

Walking around the office and talking with your employees can be a very effective management tool. Weekly, if not daily, communication helps to break down barriers that may exist between people. At the same time however, employees are not naive; they will be able to see through an insincere effort on your part. I was aware of one boss who wrote down on his calendar to "talk to the employees." This may be a good way to remind yourself to take some time to speak to the people in your office, but I wouldn't recommend it. In this example, the employees would say "it must be 3 o'clock, the boss is coming 'round to talk to us." At another office, the boss would make his tour every Friday at 4:45 p.m., and this was perceived by the employees as a check to see how many workers had left the office early. Eventually, everyone sees through this artificial behavior by a manager who is not sincere about communicating with others. Some managers have an inherent sixth sense for employee relationships. Others must work hard to achieve this skill, and some managers will never succeed at worker sensitivity.

PERSONALITY TYPES

The following descriptions of **personality types** will help you understand your employees and improve your working relationships. Over the years, I have come across a number of character traits that have helped me appreciate and understand the people I've been responsible for at work. Years later, I was introduced to the **predictive index process**, which categorizes individuals by their personality traits. I have even assigned my own nicknames to some of these mannerisms, attributes, and attitudes. It helps me to remember how I might have worked with another similar person or persons in the past. I can recall what worked and didn't work based on past experience. You can also appreciate what you would do differently, if you had the opportunity to relate to a person again under the same working conditions. None of the nicknames are meant to be derogatory. Over the years, you may develop your own list of characteristics and nicknames, which will serve as a benefit to your sensitivity and help you with your goal of getting the most out of your employees, for their benefit as well as that of the company.

The Specialist

The **specialist** is the kind of person you can count on to thoroughly complete the tasks he or she is assigned. Most specialists recognize this special trait and take pride in their work. When hiring this person, you need to recognize that you have a need for this type of individual. A list of questions directly related to the job they'll be performing can assist a manager when interviewing this type of person. For example, an experienced designer who is very good at boiler room equipment layouts will be familiar with the applicable codes, access criteria around a boiler, valving needs, etc. By developing a **job description checklist,** a manager can interview position applicants with a consistency that should provide the best candidate. On one occasion, I interviewed a specialist candidate using an equipment rooms checklist, Figure 6-1, and found the candidate was giving conflicting information. For example, the person said he was familiar with chilled water systems but not familiar with "open and closed" systems. This conflicting information allowed me to ask more specific questions regarding the candidate's experience, and I determined that he really didn't have the experience he had indicated. (Most condenser water systems are "open" systems!)

By using a checklist, you can confirm a specialist candidate's experience and knowledge. In addition, you can develop your own job description or position checklist for each position needed within the group. This also serves as a tool for the person looking to advance. The checklist provides an agenda of experience necessary to advance to the next position within the firm. The person's annual performance review can also be based partially around these parameters. There will always be other criteria to include in employee reviews.

Specialist Candidate Checklist — Equipment Rooms			
Experience	**Yes**	**No**	**Remarks**
1. Familiar with high pressure steam & PRV stations.	___	___	_____
2. Familiar with boiler feed and deaerators.	___	___	_____
3. Familiar with "closed" systems and "open" systems.	___	___	_____
4. Familiar with codes applicable to pressure vessels.	___	___	_____
5. Familiar with valve and fitting dimensions.	___	___	_____
6. Can lay out system using upper and lower level plan views.	___	___	_____
7. Can coordinate the structural weights, pads, and openings.	___	___	_____
8. Familiar with BOCA, NFPA, and ASHRAE criteria.	___	___	_____
9. Experienced at coordinating with other trades.	___	___	_____
10. Familiar with hot water systems including air control.	___	___	_____
11. Familiar with chilled water and condenser water systems.	___	___	_____
12. Can lay out floor drain and funnel drain criteria.	___	___	_____
13. Familiar with vent piping from boilers and chillers.	___	___	_____
14. Familiar with insulation criteria and material options.	___	___	_____
15. Familiar with access criteria around equipment.	___	___	_____
16. Familiar with "rules of thumb" within equipment rooms.	___	___	_____
17. Can develop a one-line diagram for each system.	___	___	_____
18. Can size pipe and sheet metal distribution.	___	___	_____
19. Understands written sequence of operation.	___	___	_____
20. Familiar with operating and maintenance manuals.	___	___	_____
21. General remarks.	___	___	_____

Figure 6-1. Example of a job description checklist for equipment rooms.

When interviewing a specialist candidate or working with a similar existing worker, a manager must recognize and appreciate what this person needs from the position. Usually, salary is not the number one issue for the specialist. More importantly, this person looks for management awareness and recognition. The drawback of having specialists on staff is their limited growth potential, because they don't see "the forest for the trees." There will be times when individuals want to advance to the next position level. Although money isn't the first priority, they will eventually reach the peak earning capacity for the position description. As they reach this financial peak, their problems become your problems. When a specialist reaches that pinnacle, he can become a financial liability to the firm.

Often times, the specialist wants to grow into a position that requires a change in character. After a period of time, it becomes obvious to the manager that this person doesn't have the skills or the personality to advance further within the company. Some people just don't change very easily. It is your responsibility to meet with the person and, over a period of time, persuade him to appreciate the things he is good at and accept the limitations he may inherently possess. I have had to do this on a couple of occasions, and it is difficult to get the individual to accept this kind of constructive criticism. If the person doesn't accept the advice and chooses to seek out another job in another company, then your problem goes away. If the person doesn't accept the advice but still chooses to remain at your firm, then you need to stay close to the issue, because performance may be compromised by the disgruntled employee. At that point, you will probably have to revisit the matter and deal with the employee's aspirations. This may eventually lead to termination.

The Theoretical Engineer

The **theoretical engineer** is very similar to the specialist and seldom sees the forest for the trees. In some areas of the hvac/r industry, this person has an integral role in the development of devices, equipment, and systems. At a research and development firm, this person will be a valuable asset, but at a design and build firm, this individual will be a liability. One valuable asset of this kind of person is that the theoretical engineer can be very specific when called upon. When designing a product or studying a problem, the theoretical engineer has the patience and perseverance to thoroughly investigate the issues. At the same time, this person doesn't work well under a time constraint. He is reluctant to take a chance, even with his wealth of experience and knowledge.

A manager must understand the benefits of this type of worker and effectively apply these talents to fulfill a job description. At the same time, you need to recognize that the theoretical engineer will slow down the design process in some "fast track" hvac/r arenas. I have found that some employees with these unique traits can never complete a job! The problem arises when you ask them if they are finished with the job and they reply that they need to finalize the calculations further. This process can go on and on until the manager steps in and makes some decisions that allow the project to move forward. In the process, you have let the theoretical engineer "off-the-hook." The theoretical engineer then divorces himself from taking any responsibility for the outcome of the project. If the project proceeds without a problem, this person often resents that you stepped in and took responsibility, because he believes he would have come to the same conclusion in time. If the project experiences a problem, no matter how insignificant, this person will make it known that you should have let him complete the task without interference. Either way, the manager loses. As the manager, you need to recognize the characteristics of this person and utilize the employee accordingly or encourage the individual to look for a job that can better use and appreciate his talents.

The Mover

The **mover** is the opposite of the specialist and the theoretical engineer. This employee clearly sees the forest but frequently misses the trees. A worker who doesn't learn the specifics associated with any hvac/r job description has very little to offer a company in the long run. Often, this type is an outgoing person and a persuasive speaker. A manager needs to assess what role this person is going to play in the company. Developing a checklist questionnaire when interviewing for a mover may be difficult to implement. Usually, these applicants have an answer for everything. Good communication skills are essential in sales or project manager positions, but the candidate must have the experience to fulfill the job function. Being able to recognize this type of worker among your existing workforce will save you a lot of heartburn later on. What I mean by this is that by recognizing the mover, you can avoid placing the person in a work role that requires attention to detail.

Over the years I have seen this personality come into a job, work for a couple of years, and then move on. When the hvac/r economy is booming, you will see these people move from job to job. With each company change, they increase their salary while building a resume of projects on which they briefly participated. For example, when a mover-type person is hired as an hvac/r engineer, he or she is assigned a project to design and is expected to participate in the project until its completion. However, nine months to a year later, if the mover becomes aware of a job opportunity at another firm, he or she may leave the first firm without seeing the design project through. The mover misses the opportunity to experience the construction phase, start-up, and close-out of the project. These integral pieces of the hvac/r business offer valuable information to a person's education. The mover, in this example, will have missed these pieces of the learning experience.

As the economy drops off and performance is monitored closely by managers, these employees find themselves unemployed. If a mover learns the specifics of a job description and stays at a firm long enough to fulfill his or her responsibilities, then this person will become invaluable to the company. There are very few employees who can excel at their job and grow into a position where it is essential that they be fundamentally sound and a very good communicator. Job security becomes an inherent part of this rare type of mover.

Not A Morning Person

I have known quite a few people who just don't like getting up for work in the morning. In particular, two individuals come to mind who were very good at what they did but did not get the recognition for their expertise. I attribute this to the fact that people took exception to having to work with these individuals each work morning. When I was responsible for the engineering and development of the construction documents for a major hospital project, Earl was the man I counted on to get the job done, and he was definitely not a morning person. I had been told by the project architect that he would not have the third floor lab drawing I needed to design the hvac/r system until a week before the deadline. In the coming weeks, I told Earl to be prepared for this late arrival and be prepared to get the job done. His job was to convert my calculations into a sheet metal distribution layout, do all the non-think work, coordinate the hvac/r with the other trades, and letter it up. When the drawing came in, Earl got the job done on time.

I remember telling my boss of my strategy and recall his amazement in the confidence I had in Earl. Because Earl wasn't the most pleasant person to have to work with in the morning, his abilities became overshadowed by his morning personality. This was an unfortunate oversight on the company's part, because the employee was a very good worker. My approach with Earl was to give him his tasks at the end of the day. This way, I didn't have to talk to him until after 10:00 a.m., and he didn't have to talk to anybody the next morning. Around 10:00 a.m., you could approach him and he would be as pleasant as anyone else in the office. This tactic contradicts my 8:00 a.m. and 1:00 p.m. review process, but there are always exceptions to the rules. For this person, planning ahead allowed the company to get the most out of him during the workday.

As the manager, you need to look beyond an individual's idiosyncrasies and focus on the job responsibilities. It is your task to get peak performance from each employee, and this sometimes means evaluating what is required to reach this goal with each worker. When hiring an "Earl," it won't be easy to recognize this trait during the hiring process. Once employed, a manager should be aware of this shortcoming in a person and make sure it doesn't create a troublesome environment for others. Over time, you could end up losing a valuable employee because both of you misunderstood the situation.

The Expert

The **expert** is another person who can create havoc in the workplace. Just as the non-morning person doesn't communicate well with co-workers in the morning, the expert can also isolate himself from the rest of his colleagues. This authoritative person will often "talk down" to the other workers because of his superiority complex. This attitude doesn't set well with the people in the office, and it is not unusual for these other people to work grudgingly with this person. It is hard to recognize the expert's "superior" attitude during the interview process. These people are often quite good at what they do, but they seem to possess a need to let you know how much they know.

Often, this superior attitude is prevalent with individuals who have more schooling and work side-by-side with engineers who have worked their way up through the ranks without college degrees. The expert has an even bigger problem working in a group where the boss is not a graduate engineer. I recall one such person who struggled to accept two co-workers, both of whom had a significant amount of hvac/r experience. Both of the co-workers were well educated, having attended numerous evening courses to improve their engineering experience. However, neither person pursued the necessary schooling required for a degree. Instead, they continually worked to improve themselves through on-the-job training. In the end, both were more recognized in the hvac/r industry for their contributions, while the expert struggled to receive recognition.

Money is very important to the expert. Experts put as much emphasis on maximum earnings as they do on letting everyone know how smart they are. A significant flaw with an expert is that **he won't tell you what he doesn't know**! What I mean by this statement is that they struggle to ask for assistance, advice, or even a second opinion. For a manager, this can be a troublesome situation, because the individual will tend to conceal project problems that he or she can't resolve. You will find it a burdensome task to spend additional time reviewing the expert's projects, because you seldom receive feedback from this person. This is especially important when the engineer has five to ten years of experience in a responsible position. This experience is still not adequate to allow him to proceed without close supervision. A manager needs to be more pro-active with periodic critiques. Communication with this individual is mandatory for project success.

The expert will remain with a firm for a number of years but eventually will move on to seek a more responsible position in another company. Many of these types of workers will strive to have their own firm, but few will achieve this goal. Instead they will procrastinate the move due to an inability to develop a strong client following. Clients usually see through this person and seldom recognize the expert to be as knowledgeable as the he wishes to be perceived.

I once knew a unique type of expert who proved to be a valuable asset. This person was a college graduate with around 25 years of experience, and he always had many answers to any question. He was a quiet individual and an excellent engineer, but he did not possess the drive to become a manager. Instead, he proved to be a valuable asset to the company when it came to training the less experienced engineers. This expert was unique in his approach to helping others. When asked a question, he would usually offer a number of alternatives. This was annoying to some of the employees, because they wanted only one solution. He tried to teach the less experienced workers that there was often more than one option, and it was up to the person in charge to determine the correct answer. He would get these less experienced engineers to recommend their own solutions to a problem. There was one "rookie" engineer who recognized and appreciated this approach. He would not hesitate to ask this expert questions, because he knew he could get at least ten good answers!

Another type of expert, the Dr. Jekyll and Mr. Hyde personality, is the person who is very good at what he or she does but is very difficult to work with on a regular basis. Some people have a personality that allows them to be very nice, almost too nice! On other occasions, they can be very abrupt and rude. I have had trouble over the years with this type of person, because you don't always know ahead of time whether they are in a good mood or a bad mood. As the manager, this is even more difficult, because reprimanding a Dr. Jekyll and Mr. Hyde personality truly tests your ability for timing. If they are not in a pleasant frame of mind when you approach them, it makes discussing the issue at hand more difficult. On other occasions, this same person will be so pleasant that you will walk away wondering whether they truly understood your message or whether they were only trying to be polite. This is not a trait that you will be able to identify very easily when interviewing the applicant.

The Graduate

Having worked my way up the corporate ladder, I know that training can develop people into excellent engineers, project managers, or service technicians. Over the years, I have hired a number of trainees, some of whom were graduate engineers. Today, I would probably hire only graduate engineers due to the sophistication of hvac/r technology. It is important that the person understands basic engineering fundamentals, such as fluids, thermal dynamics, and the refrigeration cycle. The days of the employee working his or her way up through the ranks without a college education are gone.

A graduate has a greater potential to go further in the company than someone without that higher degree of formal education. Why hire an individual who can go only so far in the company? I am not saying the graduate is automatically going to succeed, because that just isn't true. And I am also not saying that high school kids can't go to night school to further their education. The truth of the matter is that the person with the necessary credentials, who is anxious to learn and willing to invest the additional hours required to grow academically is usually the person with the best chance for advancement.

When hiring trainees, I have often told them that we are not going to pay them "top dollar" **and** teach them at the same time. In addition, I have also told new employees that if they don't progress in the next year or three years, that it will be their fault and not the company's fault. This bold statement is based on my experience working with new trainees and it's true, they do learn. In regards to the starting salary, it is important that the manager look out for the financial interests of the company. Paying "top dollar" to a trainee is not good business.

I recall one trainee who struggled with the concept for the first six months. During that time, I spoke to him on a number of occasions about his lack of effort to learn. Having just spent four years in college listening to teachers talk about how much students will earn when they graduate, this person was struggling to accept the realities of business. In addition, he had spent approximately $60,000 on his college education and was anxious to recoup his investment. I thought he was going to be the first graduate not to succeed using my philosophy. Fortunately, he recognized that the difficulties he was experiencing were due to his lack of maturity, and within the year he was on track and on schedule.

The Wanna-Be

The **wanna-be** is someone you will meet and manage quite often. These individuals always talk a good story, but they seldom achieve their goals. This type of personality is one that I recognize very quickly when given the opportunity to work with them for a period of time. Usually, it takes only a few weeks to recognize the trait. As the manager, this person will be represented in about 80% of the staff. I'm not saying that everyone is a wanna-be, but it's not unusual for people to tell you they are going to achieve a specific goal, and a year or three years later they still haven't reached the target. Managers will continually be presented with an employee's ambition to succeed as an engineer, project manager, foreman, or manager. It is your responsibility to assist employees with their professional goals. Helping individuals achieve their goals is an integral part of getting the most out of them for their benefit. Eventually, you may lose these people to other companies, because you can't always accommodate the positions they seek.

For the wanna-be personality types, ultimate success seldom comes, because they never follow through with their goals. I like to use the following phrase to illustrate this point, "Either do it or don't do it, but don't say you are going to try!" When a person says they are going to **do** something, chances are they will follow through with purpose. At the same time, people who say they are only going to **try** will most likely not succeed. When this occurs, I usually refer these people to the "bad luck" tape, which we discussed in Chapter 5. As the manager, make sure the person commits to the directive. Get them to either "do it" or not "do it," whatever "it" might be. Never accept the half-hearted commitment that they will "try." You can be certain, the goal will most likely never be met if they say they will only try.

Managers trying to recognize the characteristics associated with the wanna-bes may be able to relate to the stories of "Bill" and "Greg." Bill was someone who always wanted to be his own boss.

His goal was to have his own company someday. Interestingly, he was someone I hired as a trainee. At his job interview, he accepted the entry-level position reluctantly, because he was a college graduate and some of his classmates were being offered more money to start at their jobs. Bill bought into the advice that one year later and three years later, he would be more valuable to himself and to the company. By year's end, he could already see that he had a job satisfaction that his former classmates lacked. They still earned more money than he did, but Bill recognized and appreciated that he was continuously learning. Fourteen years later, Bill was still saying he was going to have his own company when opportunity knocked! At that time, he was approached by two different companies to join them. One company offered him the chance to be responsible for very large hvac/r project designs. These projects would inherently bring with them the status of a "big-time" engineer. The other company had a more entrepreneurial atmosphere. It was suggested that if he was ever going to be his own boss, he should take the second company's offer. I told him that the first company was similar to the firm he was already "comfortable" working in and that ten years later, he would still be telling me that he was going to have his own business. Bill chose the second firm, and within the year made the transition into a joint venture with this firm, so that he would have his own company in the second year.

Bill's performance falls into the 20% category, because he chose to stick his head above the crowd. The company I worked for ended up losing him, but his problem would have become our problem if he had chosen not to leave. The reason for this statement is that he had reached his peak earning capacity within the firm, unless he moved into management. There were no positions for new managers at that time, and this would have been discouraging to him. As a manager, helping employees to move on can allow you to maintain a staff of workers who are satisfied with their jobs. If you have a good trainee process, then there should be individuals anxious to fill the voids when other employees leave.

Another person who wanted to be in control was Greg. He was a person who clearly was in the 80% group, and he demonstrated this outlook daily. Like 80% of the employees who say they want to be in charge, Greg did little to support his goal. Instead, he was comfortable critiquing his boss's performance. As a manager, you will often have critics. This should be of little concern to you if you are in responsible charge. The goal for you is to minimize these individuals by building a staff of workers that appreciate your leadership and recognize that you are there for them, as well as for the firm. Your focus needs to be on getting the job done while looking out for both sides of the street. Greg was a good engineer but not a great engineer. He was definitely one who "tried" rather than "did." He eventually reached his peak and maintained that position for many years.

Managers should grasp the opportunity to get wanna-bes to achieve their goals. It is too easy for this type of individual to grudgingly come to work each day and be distracted by their dreams. Make them commit to their dreams, and if they don't follow through, remind them of their failure to perform. This approach isn't intended to make them feel bad when they fail to reach their plateau. Instead, it is to show them that you were counting on them to succeed. They made a promise to themselves, and they need to follow through with what they have said. In the end, you begin to minimize the idle comments directed at the company and its staff by employees who seldom backup their words with actions.

Goal setting can be an invaluable management tool for the employee and the manager to track the person's "company-oriented" goals. This will eliminate or identify wanna-be workers within the first few months. As a stipulation of employment, these goals could be used to terminate employees if they do not show signs of fulfilling the commitments they agreed to when first hired. Your target is to keep this type of person to a minimum in your group.

The Over Forty Crowd

As a person in this category, I have observed a number of people who approach the last half of their working career differently than they did the first half. Often during the work week, a manager needs to take on the role of "cheerleader." Everyone goes through cycles when they are "up" and "down" emotionally. I believe it is the person in charge who has to take the initiative of routine encouragement. This is very important when managing people. As the manager, it is important to maintain a nucleus of individuals who continually stay focused on the business at hand. Throughout the year, the workers within this nucleus may change, but you must keep an "upbeat" atmosphere within the company. To do this, you need to sense the pulse of the firm.

Within this group, you will probably have a number of employees who have worked for more than 20 years, have reached their pinnacle of success, and are now at a crossroads in their professional careers. Many of these workers recognize that they have gone as far as they are going to go in the company. Many of these people may fall into the wanna-be category and just accept their job description as it is. Some will be the mover-type personality who look at this point in their professional career as just another opportunity to continue to grow.

Individuals who accept the realities of their work and are not committed to continual educational growth, often will resign themselves to the role they have. This role can be a double-edged sword for a manager, because the person will probably be at peak earning for the position he or she occupies. At this position, the worker may not be imparting an urgency to get the job done, an excitement for the profession, or a positive outlook for the company. This person, over a period of time, will probably transform into a good but not great worker. They have the experience and knowledge to perform the job but lack the enthusiasm to express and transmit this to others within the company. What the manager ends up with is a well-paid individual who can get the job done but has little more to offer the company. Worst yet, you may have individuals who are "putting in their time" waiting for retirement. There are individuals with this attitude who still have ten to fifteen years remaining to work! Having a person like that around can have a negative impact on the other employees, as well as your clients.

On the other hand, my first boss was a person who always had a positive outlook on life. Age didn't seem to get in his way. He too started out as a trainee and eventually owned a consulting engineering firm. He was open to new ideas and expanded his firm to keep abreast of the changing hvac/r market. Although I have lost touch with him now, I am aware of the firm's activities and its success. At the same time, managers don't have to own the company to demonstrate the excitement offered in the hvac/r industry. In fact, it is a responsibility of managers to lead and not follow. You should get up every workday and enter that office with a commitment to do the best you can. At the same time, you need to be a leader who is out in front of the group providing the excitement to be the best they can be.

When interviewing candidates for employment, people with dynamic personalities can often be mistaken for having this leadership charisma. Experienced salespeople are apt to have this influence. You need to see beyond this when interviewing, because it can be mistaken for a leadership potential in a technical field. A good salesperson can be very effective at convincing you they have what it takes for the job. You need to make sure the person has the fundamental education, practical experience, and the credentials to manage. You should include these three items on the position checklist when interviewing people for a position of responsibility.

The Salesperson

Although **salespeople** are considered by some to be a necessary evil of the hvac/r industry, I have come to appreciate the really good salesperson. This is a very difficult position to fill, because the business requires detail-oriented individuals. There are very few employees who can be both outgoing and introverted enough to be able to troubleshoot an hvac/r problem. The two requirements contradict each other. Seldom will you come across an experienced service technician who is willing to speak to a group at a meeting. The same can be said for the project foreman when requested to make a presentation on an hvac/r installation. At the same time, the salesperson will often be dependent upon this service technician or foreman to assist in providing hvac/r details to a client. Most hvac/r firms can get by without a very large sales staff if their market is focused towards repeat business. A company that performs well and has a technical person with good communication skills will most likely maintain its clients. Without the salesperson on staff, a manager will need to oversee this individual and remain in touch with the client. Beyond that, the firm needs to have a person on staff who can provide marketing support, rather than sales support.

Before you hire a salesperson, the company must recognize and document the skills necessary to perform the job. This too can be recorded on a position checklist similar to the specialist checklist, which was discussed earlier in the chapter. The salesperson will be the most misunderstood and least respected employee on staff. As the manager, it may be your responsibility to oversee this individual's performance. Having a means to measure performance and minimum position requirements will be essential to managing this type of person. If it is a group of salespeople, finding a method to demonstrate their value to the company will be a benefit to your cause. Although the rest of the workers will be skeptical of this circle of people, they need to be appreciated within the company. Often, this group is talked about by the other employees because of the sales personality. Salespeople are taught to accept rejection. They are the ones who are knocking on doors, trying to generate new work. Rejection at work by co-workers is often a continuation of their job environment. A good way for a manager to keep this in-house "badgering" to a minimum is to provide visual aids that clearly demonstrate the effort of the sales force and their sales success. Your goal is to have sales performance win the respect of the workforce. In turn, it is your responsibility to make sure the workforce accepts and appreciates their co-workers and their job function.

Felix and Oscar

At some time, you can expect most of your staff to come into contact with a client. It may be over the phone, when a client visits the office, or when an employee visits the client's site. In any case, a manager needs to recognize the odd couple syndrome. You will always have "Felix" and "Oscar" characters within any company; that is, employees who dress and act very well and employees who dress and act like slobs. Employees need to understand the dress code and stay within its parameters. Each person needs to present himself or herself in a manner appropriate for the position. This means they must dress according to company guidelines and talk in a similar manner, especially with clients.

To achieve this goal, the manager may have to speak to the group, as well as particular individuals, about how they satisfy the firm's **code of standards**. A pet peeve of mine has been the person who spends hundreds of dollars on a suit and then doesn't polish his shoes. I sometimes suggest that these men consider black, ankle high sneakers with their suits, instead of the unpolished shoes. I have even given out shoe polishing kits to the employees during the holidays. Women are no exceptions when it comes to polished shoes. I've seen some women wear very professional outfits and have their shoes detract from their appearance.

Company vehicles, particularly those with the company name on it, are no different than the ankle high sneaker syndrome. Trucks with radio station bumper stickers, dirty vehicles (inside and out), etc., all detract from the company image and also make a statement about the individual inside the vehicle. The same can be said for jobsite trailers and staging areas on-site. Each of these examples present an opportunity for the employee to be classified as a Felix or an Oscar!

SUMMARY

A manager has to be a "people person." That is to say, you need to be sensitive to other people if you are to get the best performance out of them. The person in charge must understand that it is the other workers who are going to perform most of the work. You need them and are counting on them to meet their job commitment. For all of this to come together, the manager must be out in front leading the charge, i.e., the spiritual leader of the group. The more you understand about your employees' performances, goals, and idiosyncrasies, the better you will be as the manager.

Managing by walking around the office and talking with your employees offers you the opportunity to sense the mood of the office. It is your job to make sure the group is meeting its commitments to the clients. Having the sixth sense to anticipate employee attitudes will help you keep workers happy and the office upbeat. In addition to walking around, listen when the employee talks and don't interrupt. Be sensitive to his or her concerns. At the same time, maintain a certain amount of distance. Don't overstep your boundaries and become involved with the personal matters of each and every worker. Be committed to assisting them, so that their work is not compromised, but don't try to solve their personal issues.

Recognize the different personalities that are available to you when filling a job position. Different characteristics offer better results when applied to the position descriptions that make up your staffing needs. Certain jobs require certain temperaments and attitudes. A manager must be cognizant of the many types of individuals available for each hvac/r position.

Chapter 7
People in Management

This chapter is about specific people who have held responsible positions in management. I have met and worked with many of these individuals, while others I have observed from a distance. From them, I have chosen what management tools work for me. You can learn a lot from other managers. No one has yet cornered the market on great management! There are many management skills and styles to be absorbed by the new manager, as well as the experienced manager who wishes to improve his or her skills. With each personality described herein, there are management skills, methods, and traits that work and don't work. I have reaped the benefits of working with some very good and some very bad managers. My goal has been, not to dwell on the bad, but to learn how to do better.

The manager is the "hub" of the company operation and is an integral part of the success of the business plan. It is the manager's responsibility to allow the wheel to rotate smoothly, keeping the spokes strong and aligned. A department needs to be talented, perform work smoothly and on schedule, and get the job done. A project needs to proceed in a controlled environment and must be completed to the client's satisfaction. Manager performance is paramount for this process and to achieving the departmental goals. Learning from others is an excellent way to improve your management skills. Why make all the mistakes yourself?

How Many Managers Are Enough?

I have managed as many as 40 engineers, designers, and draftspeople, and I have also managed as few as eight people. If the work is consulting engineering, then that work can be very creative. Creative is another way of saying that something is not standard. As a result, I have come to believe that managing more than 12 staff members is a very difficult task. Changing technology influences the need to better; consequently, design engineering can be a perpetual motion machine. Building systems have common needs, but each job will have some features that are different from past projects. In turn, this changing technology detracts from efforts to standardize the design process. The same can be said for managers responsible for construction projects. Client requirements, timing, complexity of the installation, phasing, and certification all play an integral role in how the job is handled. Routine management in the hvac/r business is the exception to the rule.

Ideally, company design and project management standards need to be in place and adhered to, but diversity will come into play. In addition, a manager has a responsibility to maintain contact with the client, even though day-to-day distractions can hamper this client maintenance goal. Employees often complain about clients and sometimes go so far as to say "if the customer would stop calling, I could get some work done." For example, you may plan to be present at a particular jobsite every Monday at 8:00 a.m. However, at the owner's request, you and your project manager must be present at 11:00 a.m. instead. Such deviations from your pre-planned workday schedule put a burden on you to compensate for these distractions. Managing too many people, while participating in other required business activities, draws you away from your ability to manage. More repetitive work, such as manufacturing hvac/r components, gives a manager the ability to manage more people at one time. When the entire crew works in one location and produces only one product, such as fan coil units, then a manager can oversee a hundred or more employees at one time. Whether it is one manager for every 12 workers or one manager for 300 workers, there are no easy answers to the question "how many managers are enough?"

126

If you are responsible for a group of people who do not do repetitive factory work, then a good tool for determining how many people you can manage effectively is a staff and workload requirement schedule, which determines the number of hours available and the number of hours required, Figure 7-1. This may be a service group, construction crew, or an engineering team. Depending on the type of business you are in, a manager should spread the workload over a minimum of three months. This three-month scheduling projection identifies the amount of work to be performed and how many people are needed to complete the work. From this work projection, you can go back and spread your time over the next three months, including other responsibilities you may have. This is also a good scheduling tool for spreading the business plan/budgeted sales manpower projection. With this simple scheduling process, you can measure "budget hours" to "actual hours." With each passing year, you will become better at managing workload and staff requirements. At the same time, you can determine how many hours in the week you will have to spend managing the process. How many managers are enough? You can determine that by performing some simple math.

My First Managers

When I first started in the hvac/r industry, I had the opportunity to work with an excellent business owner/engineer and an equally talented chief engineer. Both individuals provided me with many of the building blocks that have continued to support my efforts today. Their organizational skills helped to shape my management style. At the same time, the on-the-job training offered me the opportunity to be exposed to and part of an up and coming consulting firm. The environment provided me with the chance to excel in many areas of the hvac/r business. This first job also introduced me to the pressures of deadlines and client commitments, as well as the routine of getting coffee, making prints, and delivering documents to architects and jobsites. In the eight years I worked for this company, the atmosphere was always "if you can do two jobs then we'll give you three jobs."

Staff and Workload Requirements

Activities	April	May	June
Budgeted Sales			
Project A: 900 hrs required over six months (Start-April)	100	130	400
Project B: 950 hrs required over four months (Start-April)	140	300	400
Project C: 400 hrs required over six months (Start-June)	0	0	100
Budgeted hours required	240	430	900
Booked Sales			
Project 1: 1240 hrs required over six months (Started-March)	400	200	200
Project 2: 900 hrs required over five months (Started-April)	400	300	300
Project 3: 600 hrs required over two months (Started-July)	0	0	0
Booked hours required	800	500	500
Other Activities			
Meetings: 3 hrs per person per week (6 people)	72	72	90
Training and quality control: 12 hrs per week	48	48	60
Sick: 8 hrs every other week	16	16	20
Other hours required	136	136	170
Total hours required	1176	1066	1570
Available Staff			
Senior engineer: Bob	160	160	200
Engineers: Ned and Stu	320	320	400
Design engineers: Herb and Jim	320	320	400
CAD operator: Kofi	160	160	200
Available hours	960	960	1200
Summary	April	May	June
Total hours required	1176	1066	1570
Available hours	960	960	1200
Surplus	—	—	—
Deficiency	(216)	(106)	(370)

Three-month tactical plan comments
- Allocate ( ) surplus hours to archiving job folders/drawings.
- Implement a (10) hour workday to accommodate deficiency hours in April.
- Outsource CAD work to accommodate deficiency hours in April and June.
- Based on budget sales staying on target, hire another senior engineer by mid-May.

Figure 7-1. Example of a staff and workload requirement schedule.

It was at my first place of employment that I was introduced to such management tools as time management, "things to do" lists, action agendas, standardization, and writing letters. All of these tools have served me well through the years. The concept of bringing along a trainee and spending time teaching this person began when I was hired. The concept works, and it too is an invaluable management tool for me today.

Both managers were good engineers who could communicate well with others. Each had his own distinct skills, which complemented the other's abilities. One was proficient in sales and negotiating, while the other was adept at details and production. Together, I was able to absorb many of their tricks of the trade. Back then, management focused on speed and accuracy based on the art of free-hand sketches and the use of 8-1/2" x 11" paper. When designing an hvac/r system, you were expected to sketch it on a standard sheet of paper. Wasting time by drawing bigger pictures was not cost effective. Time management was the cornerstone of their philosophy. Being able to "see the forest for the trees" was paramount to anything you were assigned to do. This concept also applied to managing people. I was taught to see the big picture. Sensitivity to your work, the work of others, and to the heartbeat of the office were by-products of this concept.

The Loyal Boss

In my experience, I have found that some managers are too loyal to the company, while others are too loyal to the employees. Always remember that management is a two-way street! A manager has to look out for the company **and** the employees. Loyalty doesn't have a place in the management tools of a manager. This job responsibility is a fine line to walk. You need to make decisions based on a sound business plan and not on personalities. These decisions need to be correct and fair and without bias or prejudice. Established managers often function "with blinders on," always taking the company's point of view. First-time managers, on the other hand, tend to take the employees' side in a dispute, because they were recently on that side of the street themselves. In time, this new person will begin to lean towards and agree with upper management's point of view.

Frequently, a manager must make a choice between what is popular and what is in the best interest of the company. This was very obvious to me on two similar occasions. In both examples, I was assigned additional management responsibilities that included more employees joining the group. When taking charge of these added individuals, I reviewed their personnel records and determined that two workers were being highly compensated for the jobs they performed. Their salaries were significantly higher than what the job responsibilities required. It was obvious to me that the last manager had not paid attention to detail or was not able to confront the individuals about their salaries. Their problem became my problem. A manager must continually monitor the performance of the workforce. It is easy to give out salary increases and bonuses, but the manager must also be the "bearer of bad news." You need to be prepared to fulfill all the job functions of a manager. This includes not overcompensating workers for the work they perform.

In this example, the company had been overcompensating individuals for the positions they occupied. The employee-loyal manager may not have wanted to upset the individuals by investigating what the value actually was for the jobs they were performing. It is often said that it is a manager's responsibility to spend sufficient time prior to meeting with an employee to determine what is fair compensation for his or her services. A loyal boss needs to be loyal to the company, as well as the employees. **Do your homework!** Know what the marketplace is paying for each position you have in your group. Remove the loyalty factor from the agenda by not arbitrarily allocating an annual percentage increase to the individual's salary. This is particularly important when assessing a trainee's performance and the experienced specialist's performance. In the case of the trainee, the fair value increase may be significant and easy for a loyal boss to announce to the individual. In the case of the experienced specialist, the person may be at his or her peak earning capacity, and a cost-of-living increase may be the only increase given. This is where a loyal boss will have trouble broaching the issue.

In the examples above, I was put in the position of being the "bad guy" when assigned these additional company responsibilities. I inherited employees who were overcompensated for their job functions. At the same time, there were other workers who were significantly underpaid for the positions they held. As the manager, you need to be able to make the hard decisions, even if you didn't create the problem! This can mean putting a freeze on a person's salary, cutting a person's salary, or moving in the direction of terminating his or her employment. Any of these corrective measures are difficult to implement, but the manager must do what is fair. Being a loyal boss serves no useful purpose in the long run.

The manager before me mismanaged this issue when working with these highly compensated workers. In one case, I believe the manager didn't like to be placed in the position of being the "bad guy" by not giving the employee a salary increase. As a result, the employee's salary increased over a period of years, but his level of effort, performance, and knowledge of the business remained the same! In the second case, the previous manager failed to evaluate the person's performance properly. Instead of truly measuring the individual's achievements, the manager provided a salary increase based on his concern for losing the person to another company. At the time, business was very good, and this manager believed he needed to increase the employee's salary or risk losing him to the competition. This logic has no place in the evaluation of someone's performance. A manager must put aside the desire to be a "good guy" and manage fairly.

GOOD, BETTER, AND BEST MANAGERS

There are a lot of "good" managers, and there are a lot of managers who are even "better." But unfortunately, I can count the number of "best" managers on one hand. In addition to the bosses I have had, I am aware of only a few others who are in what I believe to be the best manager category. My knowledge of these other individuals has come through networking with others about how they manage. Unfortunately, there are no manager groups that meet quarterly to discuss and share their experiences as managers. Such a group could focus on the issues related to managing more effectively based on their experience of what works and what doesn't work. Instead what I have done is query people in other firms about their companies and how they function; the companies' good points and bad points; and employee stability.

I have known a number of "good" managers who didn't truly comprehend their employees' skills and value to the company. Each of these managers was friendly to their employees and would socialize with them on occasion. What the "good" manager lacks is the sensitivity to understand people and the ability to help those people be the very best they can be. This insight into what would be best for the employee now, in a year, and in three years is a managerial trait that the "good" manager doesn't comprehend. This sixth sense is a skill that only the "best" managers possess. A goal of every manager should be to develop the skills necessary to inspire people to want to work. These skills will take time to develop, but they can be achieved by any person sincere about the performance of the company and the individuals within the company. In time, people can sense that your goal is to have everyone succeed, not just a select few individuals.

The "better" manager, is someone who shows signs of brilliance but lacks the consistency to maintain the level of effort needed to continually manage by walking around and talking with employees. It is a lot easier for managers not to ask about the employees' goals, needs, and concerns and not to continuously encourage them to **do** rather than **try**. This type of leader needs to overcome the assumption that managers don't need to constantly encourage their employees to do better. However, where the "good" manager will not put in the time necessary to inspire workers on a regular basis, the "better" manager will. At the same time, it is difficult to continually strive to be the very "best" manager. When you first take on the role of a boss, the excitement, challenge, and opportunities are obvious to you. Three years later, the excitement may begin to dwindle, and six or seven years later, the challenge has become a routine! The "better" managers walk a fine line between routine and exciting. Where the "best" managers can continually find a new, improved way to lead, the other managers may stumble. If you are going to be the best you can be, you need to get up every day with an excitement to excel.

Another characteristic of the "best" managers is that they are able to maintain a nucleus of workers for a minimum of five years. This is a reasonable milestone to strive for and meet. Reducing the employee turnover within a firm, whether it is a service, construction, or consulting business, should be a priority of any manager. Being in charge of a staff that changes personnel every two or three years is not good for the company. In addition to sending out a negative signal to the competition and to the clients, it is just bad management! The reliability of the people within a company is an integral part of how that company is perceived in the hvac/r industry. This "revolving door" image has been given to a number of businesses in the city where I currently work. This image reflects back on the company, but it really should continue on down to the manager. It is the manager who can effectively control the speed of this revolving door. The "best" managers will have legions of admirers who not only enjoy working with the manager but will openly speak up for the person.

Often, a manager doesn't want to become too involved with the employees, thinking that they will become close friends. A manager needs to manage at arm's distance, but that doesn't mean they shouldn't be supportive of the workers. It is good to socialize with people on a selective basis, although I discourage managers from participating in routine gatherings. These meetings often turn to company talk and how the company may or may not be doing enough. Employees should realize that the manager has a responsibility to look out for the company, as well as the workers. Employees seldom take the management's point of view on a continual basis. As a result, these same people seldom appreciate your presence at a gathering of co-workers outside the office. Managers should selectively seek out the general consensus without becoming closely involved with the group. The same can be said for the manager who chooses not to manage by walking around and talking with the employees. This concept doesn't always have to be in the office. It can occur outside the office with a group of workers. Managers who don't seek out the workers' opinions will never become the "best" managers they can possibly be, because they will not be getting the valuable feedback they need from employees on a regular basis.

THE WRONG LEADERS

In order to be a very good manager, it helps to work with others who are **not** good at management. Over the years, you will come across far more managers who aren't good at managing than those who are. When working for or with a person with marginal leadership skills, you should take note of what they do wrong. Whether you address these issues with the person is up to you. More importantly, learn from their mistakes! I have broken down "bad" managers into three categories.

Mr. Big

The skills this type of boss possesses, which you don't want to have, are related to image. He is someone who has made it into management and believes that finally the company has recognized how valuable a person he is. It is not unusual for some people to get caught up in the status of being in control. As the person in charge, you need to establish your authority to give direction and earn the respect of the workers taking this direction. How you present that leadership will affect how well the employees perform. Remember, you need them as much as they need you.

I have heard many stories about the Mr. Bigs of the hvac/r industry, because many of the managers today were service technicians, designers, and maintenance technicians not so long ago. I was made aware of one particular technician who had worked his way up the ladder of success into a corporate management level position. In the transition, he seemed to have forgotten who his co-workers were. Once a week he would remind these people that he was the "senior manager" of maintenance. It was obvious to everyone that he had been promoted, so he didn't have to keep reminding his co-workers. Instead of impressing the employees with his achievement, he seemed to be insecure with his position. A person is expected to perform when promoted up the ladder of success. Through goal setting and close measurement of those goals, your actions will speak for you, and your employees will recognize that you are getting the job done.

Mr. Useless

These individuals simply amaze me, because it seems so apparent that they should never have been promoted. You will come across a select few individuals who fall into this category. However illogical it may seem, some people make it into management despite their uselessness. I have come across one or two of these people and have found that there is nothing to learn from them, not even how **not** to do something. What is even more amazing is that they seem to survive in a management role for many years.

The one thing you do want to learn from a person with limited skills is to not put them into a position of responsibility. Put aside personalities and friendships when it comes to promotions. Whether it is a lead technician, foreman, project manager, etc., maintain a focus on the job description and the person's ability and skills to succeed in the new position. When presented the facts, it is better to have a friend be angry with you because of a lost promotion than to promote an unqualified person. You will have to live with your mistake and work with Mr. Useless daily until you can't afford to have this person continue to be in charge. At that time, if the person was a friend of yours, he or she most likely won't be when you terminate them. And by that time, termination will probably be your only option, because it is very difficult to take back a management role once you have given it to someone.

A more subtle Mr. Useless is the person who has been both a loyal employee and very good at what he or she does. Other people may call this the Peter Principle which was discussed in Chapter 3. This person is more common in management and should be avoided. There have been numerous individuals who have worked their way up the corporate ladder to a point of incompetence. The word "incompetence" may be a little strong, because Mr. Useless may be doing a fair job of managing; however, he is not doing a **great** job of managing. This person's effectiveness will hurt the company in the long run, because he is not the best person for the job. Again, it is very difficult to take something away, particularly if this individual was successful for part of the time and was a good and loyal company person.

As a manager, never allow company loyalty to be a prerequisite for promotion, and always make sure the individual has the skills for the job before giving the promotion. If their skills are in question, implement a trainee agenda for a specific period of time. Depending on the new position, the time can be set and agreed upon by both parties. Also, clear goals and a means of measurement are needed. The worst thing a manager can do to these individuals is allow them to give up what they are very good at, to perform a job that is not appropriate for them. You not only sentence them to a job where they can fail, you will lose a person who truly likes to work at your firm.

Mr. Can't Do It

I believe there is nothing worse in management than this type of manager. Not to be mistaken with being a "yes man," a manager must always be positive and determined to succeed. It is the manager's responsibility to direct and reach the goal, project deadline, solution, etc. Leading the charge from the rear is strategically not a very good management decision, and this is what the Mr. Can't Do It type of person does. The manager who consistently says "it can't be done" needs to be replaced by someone who will find a way to get it done. You can be assured the competition will gladly get it done for you!

Over the years, you will come across pessimists who want to be in charge but always look at things in a negative way. Having worked with a few of these managers, I find it very frustrating to be on the same team with them. With these people, it seems as if there is always a ritual of finding the reasons why something can't be done. After all the excuses are presented, these people will then begin to work and get the project done on time. It was their ability to perform that first got them into the role of manager. Once in that position, there appears to be an insecurity that needs to be stated. By constantly saying "it can't be done" and then completing the work, Mr. Can't Do It has shown that he can do the impossible. Unfortunately, very few people appreciate the end product, because they knew he could do it; he always does!

As a manager, the words "can't do it" don't belong in your vocabulary. Instead, learn to be creative. Solve the dilemma by scanning all your options. Can people work overtime? Can you farm the work out to subcontractors? Can you get the basic system operating and make provisions to phase in the other pieces at a later date? Although some things are impossible to achieve, most things **can** be accomplished. I remember one boss who was competing for an engineering design contract against another firm. It had come down to two consulting companies, and the project schedule required the engineers to be finished in 12 weeks. The other firm had indicated that 14 weeks was the absolute minimum time needed to complete the design. The person I knew didn't disagree with this statement but said he would work within the architect's project schedule. He got the job and finished it in 12 weeks, plus two weeks during the addendum period, for a total of 14 weeks. He knew there were four weeks scheduled for bidding, and he would have at least two weeks during that time to issue an addendum for changes in the project documents. His firm finished the project in 14 weeks, just like the other engineer said it would take to complete the job.

DOUBLE STANDARDS

When a person reaches the management level, he or she often gets consumed by the position. Unless you own the company, you need to recognize that you are still an employee of the firm. **Rank does not have its privileges!** Managers must work by the same corporate rules as everyone else within the firm. It is very easy to justify to yourself that you are entitled to special treatment, special benefits, and carte blanche on your expenses. You might consider the extra time you put in at home, your work on the monthly manager's report, or the time away from home when you had to take a business trip. These are just a few of the misappropriations that go on within the management ranks. It is better to identify exactly what is allowed above and beyond your annual salary. You should identify the fringe benefits that go with the position when you take the job. Don't leave anything to interpretation.

Some managers will continually justify coming in to work late by citing that they worked until midnight the day before. At the same time, most people in the firm are probably aware of the fact that this manager is routinely late for work. Another example of position abuse is the "do as I say, not as I do" syndrome. Extended lunches and leaving early can become habit-forming. Issuing a memo to your staff that they need to be to work on time, even though you are not, can be an annoyance to the workers. These daily contradictions, that can be seen by the employees, send a negative signal to the workforce. You need to set the example. Just like everything else you do as a manager, establishing a single standard for the workers to work within is essential to a good working relationship.

Often, managers will forget that they are responsible for setting the example. Instead, they may opt to "let up" on their performance. When this happens, workers usually are quick to notice. This is particularly so with disgruntled employees. You can never allow yourself to work at a level less than what you would expect the others to function at. The person in charge has to set the pace. If you think back to people you had as your boss, what was your reaction when they arrived to work late; spent a considerable amount of time discussing their golf game with another manager; or left in the middle of the day? What did you think when the boss arrived late to a meeting? If that had been you who was late, things would have been different. Your boss would have probably commented upon your late arrival.

Another issue with people in management is **timing**. I recall one firm that had 16 vice presidents, each with a company car! There were approximately 160 employees in the firm, which equated to one vice president for every ten people! You didn't have to be a rocket scientist to figure out that the company had a surplus of management. As business began to slow down, I observed this company beginning to feel the economic effects of this drop in the economy. This down turn in the business didn't seem to affect these numerous vice presidents. A number of them had lost focus of their responsibilities and were caught up in maintaining their fringe benefits. In time, most of these people were terminated or encouraged to leave. Had they collectively concentrated on the business issues and problems at hand, and not their own territorial issues, I believe they could have had a positive impact on the company's business plan.

When you have worked very hard to reach an occupational plateau, you can easily fall into the thinking that you've finally made it! A manager can never sit back and coast, because others are counting on you and watching what you do. When business is projected to be slow for an extended period of time, then it may be wise to keep that company vehicle for a year longer than the leasing agreement. When you finally decide to turn the vehicle in, maybe you should consider a less expensive automobile. You need to know how to time your privilege breaks. Be discreet with that combination business trip/vacation or new office furniture. Rank truly doesn't have its privileges in the hvac/r industry.

ESTABLISH YOURSELF

Aggressive people who move into management positions often will make the mistake of not **establishing themselves**. The urge to move to the next level can cloud the person's management plan. When I moved into a management position, I strategically and tactically laid out a three-year plan. Often, managers will overlook the accomplishment of reaching the level they currently are occupying. Within a year, they are anxiously thinking and planning their next promotion. These people forget that they have a responsibility to perform at the job they are at and continuously succeed at that position. Instead, they have set their sights on a new role with more responsibility, authority, and money.

I know of an individual who joined a service company and within a year had the good fortune of being part of the company's increased sales. It was my understanding that this person had arrived at a time when a select few clients were on the verge of signing large contracts with this company. I do believe he was instrumental in helping to close these contracts, but months earlier, others were hard at work beginning the closing process. This new manager didn't recognize the efforts of key individuals within the firm who had worked hard on these service accounts for a number of months.

139

What clouded this manager's assessment of the closing was that he failed to give credit where credit was do. Instead, he was reaping the benefits of these closings, partly because he was now in charge and partly because the contracts were signed after his arrival. Immersed in the accolades, he was an overnight success! Before the year was out, he was focusing his sights on a vice president position, the number two position in the company.

Although management didn't recognize this person's new focus, other people within the company could see where his priorities were now headed. Because he had received most of the recognition for this new work, there were others who took exception to his plans. As a manager, you need to make sure that the entire team receives the credit for a job well done. Rest assured that as the person in charge you will receive your credit. Remember, these people are in your group, and the group's successes reflect on your ability to maintain a good staff. You can contribute too, but for the most part, it is the other person's job responsibility to bring in those new contracts. Your job is to manage the process, assist when needed, and provide overall guidance. In the case of this particular manager, he unfortunately deduced that the closings were a direct result of his late input. After these triumphant first-year achievements, he didn't consider repeating his successful performance. Instead, he was anxious to grab the next brass ring.

The problem with being an immediate success is that you have to ask yourself "what do I do now to continue being the leader everyone thinks I am?" Having been a manager at age 31, I was conscious of this other person's impending dilemma. I recognized early on that when you are a success early in you career, you inherently consider moving up to the next position. In the hvac/r industry, there are only so many positions available and only so much money to be made. Each job function has a salary range, and a manager can go only so far within a company. True, there are exceptions to the rule, but unless you own the firm, your salary and position of authority will reach a cap.

You need to assess your progress when moving into management. The problem with being the number two or number three person in a company at age 31 is what you are going to do for the next 31 years? The same can be said for nearing the top of your peak earning potential. Whether you are 31 or 41, you still have a long time to manage. Managing the same department for the next 20 years may create some heartburn when you consider that you have only worked for about 20 years to date! You have plenty of time to establish yourself. Don't be anxious to rush to the next position, because there are very few job openings once you have reached management level.

SUMMARY

As the saying goes, "good guys finish last!" However, people in management shouldn't be good guys or bad guys. A manager needs to be fair, profitable, and a person who can solve problems, both personal and professional. You can't be all things to all people. Consider your position as the "hub" of the group from which all things revolve. Maintain a smooth rotation, balancing workload with staff requirements. Be proficient in time management. Maintain a "things to do" list, and record all your meetings in an action agenda. These essential management tools will provide you with the ability to keep the wheel turning smoothly and on schedule. Certainly, standardization will also help a manager accomplish these tasks.

A manager must work at focusing on the issues at hand when it comes to employees. Anyone can make the popular decisions, but very few managers can make the unpopular resolutions. You need to divorce yourself from the group and look at issues from a perspective that is in the best interest of the company and the individuals within this company. Loyalty, in lieu of good management sense, will only get you into trouble further down the road. Problems don't go away unless you resolve them. Managers will address tough questions differently, and how they resolve these questions will distinguish them as good, better, or best managers.

Don't try to do everything right yourself. Learn from the managers you have had the opportunity to work with in the past. Grasp the opportunity to repeat the things they did, which you considered to be effective. At the same time, learn what not to do from those managers who were not so good. A person trying to be the best manager he or she can be can learn a lot from his or her counterparts in the business. Ask questions of other managers, listen to what they have to say, and learn from their experience.

Be an opportunist, but stick to your commitments until you have succeeded. Don't be anxious to move quickly to the next opportunity without finishing the business at hand. Too often, an aggressive manager will take on a responsibility and not see it through to the end. Leaving unfinished business to someone else and believing you did your part is an error. It is a tempting way to get ahead, but the best managers will follow through with their plans and repeat their success before moving on to a new challenge. Be persistent. Avoid the temptation to grasp the next brass ring. A leader must see a commitment to its successful completion. Employees will recognize, respect, and appreciate your obligation to perform your job and will resent a manager's self-centered goal to get to the top. Establish yourself and your management plan before you move on. Don't send a double standard signal to the workers. Set the example, be a positive person, and get the job done.

Chapter 8
Employee Performance Reviews

During the year, employees have hourly, daily, weekly, and monthly commitments to the company for which they work. No matter what their job descriptions are, employees have responsibilities that must be met when performing certain tasks. These commitments usually have deadline requirements that coincide with the completion of a job. Meeting deadlines is an important responsibility and part of a person's obligation to fulfilling his or her job description. On the other side of the street, the employer has very few deadlines they are required to meet to satisfy the employee. A company does have various obligations and commitments to its employees, but most of these issues are not tied to a specific time period, deadline, or individual. However, the one deadline the employer does owe the employees each year is their annual performance reviews.

Completing the employees' annual performance reviews on time is one of the most sensitive employer obligations. Whether the review is for an individual or for a union group, the deadline is known months, if not years, ahead of time. During the period leading up to this date, employees have ample time to ponder their upcoming review anxiously. There may be no other single company issue that can be as demoralizing to a person as a late review and/or renewal of a contract. Yet many employers often let this anniversary meeting slide and be missed. On the other hand, staying ahead of the personnel reviews with efficient, pro-active action sends a signal to the employees that the company values them and is committed to meeting its responsibilities on time.

When I first moved into my management role, the company had a policy of providing personnel reviews twice a year. These evaluations were implemented just prior to the dates they took effect, January 1st and July 1st. An advantage to this approach was that it gave the employee and the employer the opportunity to meet twice a year to discuss personal progress. A disadvantage was that it did not provide management with sufficient time to adequately implement the review process. Depending on the size of the firm, the manager probably was not able to spend the time needed to properly communicate and explain the employer/employee evaluation. This was a missed opportunity for the manager. As the new manager of engineering, I carried on the policy of reviewing people twice annually, but the number of people within my group confirmed my observations. The concept was less effective, because the large number of people was an obstacle to its success. The reviews had to be shortened so they all could be completed on schedule.

Prior to being the manager, I was skeptical of the review process because it was often brief. You were told how much more money you would be earning, and maybe there was a brief discussion about what you were doing well and where you could improve. It was not obvious to me, until I had to do the reviewing, that there were too many employees to make the review process a success. I thought more time should be spent discussing the goals of the company and the individual, highlighting their achievements, noting the areas of needed improvement, and setting new goals for the coming year. My department had 34 people when I became the manager. I quickly realized that there weren't enough days in the week for me to complete the interviews the way I wanted. Thirty-four reviews, multiplied by two per year, equaled 68 reviews in a 50 week period (assuming I took a two week vacation). It would have been impractical to implement my concept to improve the employee reviews, because I would not have had the time to fulfill my other responsibilities sufficiently. Before my first management year was over, I successfully lobbied to replace the semi-annual review with one comprehensive interview annually.

A benefit of semi-annual reviews that is worth noting is that the employee usually received a pay increase twice a year. In a booming economy, I noticed that a person would receive a salary increase, and it would take about three months before the person was comfortable with the new salary. After that, workers often began to think about how much more money they could be earning. This was quite prevalent in the high tech boom years, and I observed it firsthand with some of my friends who were specialists in the computer industry. It wasn't unusual for one of these specialists to change companies every 18 months to two years. The semi-annual review helped to control this problem, because the employer was addressing the employee's financial needs in a timely manner. The two semi-annual increases were equal to what one could expect with an annual salary increase, but the employee was receiving a raise every six months.

When changing the policy and procedure, I recognized the employee review needed to take effect on the person's date of employment. I had a number of reasons for this. First, this approach spread the annual review dates randomly throughout the year. This offered me the opportunity to spread the meetings with my 34 employees over a 50-week year. A second reason is that if the company changed to one review per year but stayed with the reviews taking effect either January 1st or July 1st, I would still have to meet with approximately half the employees within a short period of time. This was exactly what I was trying to get away from. Third, when all or half the employees are scheduled to have a salary adjustment, expectations and company gossip can run rampant as the due date approaches. Spreading the meetings over the full year eliminated this problem. A fourth concern with all the reviews taking effect January 1st was the possibility of people leaving the firm soon after. An observation I had prior to joining management was that January 1st was the optimum time to leave a company. The fiscal year was over, and if the company had a policy of giving bonuses at the end of the year, staying until December 31st made economic sense. Leaving the first of the year provided the employee with his or her bonus and time to accumulate vacation hours prior to the next summer season. As workers approach late spring, leaving becomes less attractive to them because of their upcoming vacation plans. The second review/ salary increase occurred July 1st, and scheduled vacations seemed to take people's minds off any complaints they might have in the workplace.

I also noticed that the next time period when workers were most likely to look elsewhere was at the end of the summer. Vacations were over, and the employees had two months to settle in before the holidays. Seldom did anyone ever change jobs between November 1st and January 1st. Over the years, I have monitored these benchmark periods, because they provide important signs of employee satisfaction and dissatisfaction. If a worker is not happy with his or her job, it may be a problem the manager can solve if sensitive to the workplace atmosphere. If you forget an employee's review, you send a signal that this issue is not important to you. Completing the evaluation on time can be interpreted by the employee to mean that they are considered a valued worker and the company is committed to meeting its responsibilities.

MANAGING BY WALKING AROUND

In replacing the semi-annual reviews with one annual review, I also learned to **manage by walking around**. This is a phrase I became familiar with only in recent years, although I have always made a point of walking around talking to employees daily. Getting out of your office and moving about is mandatory for a manager to be effective. You may have observed that some managers are more comfortable concentrating on their own work within their own office. As the manager, you need to be a leader. In order to be in charge and in control, management cannot lose touch with the team. Being preoccupied with your daily work schedule and not staying in touch with the employees is a common mistake. If you don't get out of your office and communicate with the workers you will lose touch with them. Whether the company is small or large, managers must be out talking to the employees. Project status, client issues, commitments, and the general pulse of the company morale are essential topics of conversation. Managers seldom have the time or the responsibility to commit 100% of their schedules to any one project. The manager usually monitors the progress of multiple jobs, and it is the workers who are getting these jobs done. Staying in touch provides you with daily status reports without having to hold a formal meeting to find out where the project stands.

Networking within the office provides the manager the opportunity to randomly monitor the progress of each and every worker. Through frequent discussions during the year, the manager should be aware of each worker's goals and the means they are taking to achieve these goals. This helps the manager ensure that each worker achieves his or her milestones. This method of management offers a daily review of the employees. Via constant communication, employees will know where they stand with their success and progress prior to their annual reviews. These same daily/weekly conversations demonstrate the company's interest in the worker. From a business point of view, daily or weekly communication is a valuable tool to monitor the progress of each job. By staying close to the people responsible for the project, the manager stays close to the project itself. Lose touch with the team and management will lose touch with success!

Walking around also provides you with the opportunity to catch someone doing something right. This tool, as outlined in the book *The One Minute Manager*[1], offers you the chance to discuss a topic with a person to determine if he or she is on top of the task. If the employee is not the engineer in charge, you still have the opportunity to get another person's opinion as to how the project and/or work is going. No one is going to come into your office on a regular basis and provide you with this information. You've got to get out there and hear it firsthand. Another *One Minute Manager* benefit to managing by walking around is that you don't have to wait until an employee's performance review to correct something that is being done incorrectly.

Two Minutes or Two Hours — The Employee's Review

I have developed the following review format within my management guidelines, and I believe in it completely. The employee review gives the manager and the worker the opportunity to communicate one-on-one. It is the one opportunity annually for the employee to be the focus of the company. Often, the people in charge lose sight of the teamwork needed to provide the product.

Just as often, these leaders will lack the consistency needed to be the best they can be at their jobs. Employees look to management to set the pace. Managers should strive to maintain, and hopefully accelerate, the pace at a constant rate. Through daily reviews and the annual review, workers know that management is committed to their success.

Prior to meeting with employees at their annual review, you should let them know that you need to meet with them and that they must be prepared to talk about their goals, etc. This must be in advance of the annual review date. Prepare for these meetings by reviewing the individual's past performance and making note of specific job performance and past pertinent conversations. In addition, you should know ahead of time exactly how much of a salary increase you will be recommending for the employee to upper management. People always want more money, but money isn't everything to everyone, every time. Most employees want to be happy in their jobs. They may want advancement, challenge, job satisfaction, direction, and/or a need to be part of something. As the manager, you need to let them take center stage and give them the time to express their points of view. Whatever time is necessary to complete the employee review must be spent at least once a year, whether it is two minutes or two hours.

Time management is also important when implementing this strategy. You should come to the meeting with an agenda, as should the worker. You need to know as much about this person's work and personality as you can in order to successfully satisfy your agenda and the employee's agenda. A manager who is not close to the pulse of the company may find this time commitment a problem to implement. This type of review won't work if the manager hasn't spent the past year communicating with the staff. Managing by walking around will be invaluable to you in regards to this aspect of preparation.

A good set of notes, taken during the review meeting, can avoid any misunderstandings later. Any outstanding issues should be documented and completion dates agreed upon. If the worker raises an issue that you are not prepared to answer, make sure you state a time when you will get back to this person with an answer.

Don't leave the meeting with any unfinished business, and don't let any interruptions occur during the meeting. The sole purpose of this "get together" is to meet with a valued employee to discuss his or her performance. Interruption should be avoided and prevented.

The Interview Agenda

A well planned interview agenda is essential to the success of the review meeting. An agenda is essential to those meetings in which employees have the opportunity to participate and have their say, whether the meeting is two minutes or two hours. Prior to meeting with an employee, a manager should prepare an agenda to assist in the annual review process and to avoid losing control of the meeting. Otherwise, the meeting could turn into a gripe session where nothing is achieved. No meeting should occur without an agenda. It goes without saying that a project meeting should have an agenda, but an employee review should also have one.

Each employee should already have a set of goals that were documented from the past year's review. Goal setting allows employees to create a road map of what they plan to achieve in the coming year. These milestones should be company-oriented instead of personal, and management should have participated in their preparation and given approval. Personal goals, unrelated to work, should be set by the individual outside of the workplace. The intent is to focus on the firm's goals, employee participation in reaching these goals, and employee job satisfaction. These goals also challenge employees to improve and be the best they can be at the jobs they do. Often, this results in the workers recognizing that the company values their performance. These goals also offer a means by which the company can measure the workers' performances.

The review is a valuable tool for the manager to evaluate employee performance; just as importantly, it is a tool for the employee to evaluate his or her own success. Together, it becomes very difficult for a person to leave a meeting in disagreement as to the contents and results of the review. The two minute or two hour review process, when used with an agenda, provides a controlled forum where issues can be discussed one-on-one. Again, note that this meeting should be scheduled in a location that allows the participants to complete the process without interruptions or time restraints.

FAIR VALUE SALARY

An example I draw upon, particularly with younger employees, is my own story of how I started in business. When I graduated from high school, I took a one-year drafting course, because I wanted to learn how to draw and wasn't interested in college at that time. Drafting offered me my first experience with job satisfaction. Starting at $75 a week, I recognized early in my career that challenge, job satisfaction, direction, and participation in the success of the company were very important to me. My outlook towards work was that I was receiving $59.75 a week (after taxes), and I was learning something each and every workday. If I had gone to college, I would have been paying to learn. Instead, no matter how much (or how little) I earned in my first four years, at least the money was coming in, not going out!

Managers who are preoccupied with the financial aspect of a review can underestimate the individual being interviewed. Most employees I have worked with want more than just money; they want to be a participant! This is an important issue when sitting in with employees during their reviews. On many occasions, I have said to employees, "we want to get the most out of you, for your benefit as well as the company's benefit." I point out to them that they should look at where they are going to be in one year and three years. If they don't learn their jobs sufficiently in that period of time, I tell them that they have only themselves to blame. I can make this statement with confidence, because I have been there before and the process has worked. Based on my experience and method of management, I **expect** them to get ahead, and I am confident they will get the opportunity to excel!

In my earlier years, I continually felt I was being "beat" for about 10% of my true value to the companies I worked for. However, I had the maturity to recognize that I was constantly learning, and in the long run, I was getting ahead. Years later I can say that I chose the correct path to success, both financially and educationally. This career course needs to be expressed and understood by employees who are anxious to succeed. As the manager, it is important to stress the combination of earning a fair value salary while being taught by the company. It is not practical for a company to pay top dollar to an employee if the company is also teaching that individual. The two-way street philosophy dictates that employees being reviewed appreciate that the company is investing in them both financially and educationally. If employees truly want to get ahead, they must understand that there is nothing wrong with a company paying them a fair value salary for their services. They should also recognize that they probably take longer to complete a task than a more experienced, higher paid worker. It is the manager's responsibility to reinforce this philosophy in the review meeting, i.e., **the company is not going to pay you top dollar and teach you at the same time**.

Most people in management worked their way up through the ranks to some extent. Some managers may have started in the middle, as in the example of an engineer-in-training. This person may not have started at the lowest position in the company (i.e., draftsperson), but still they progressed through an educational process that got them to where they are now in the organization. A manager should reflect back on his or her experience and try to understand where the employees are coming from. Often, I've noticed managers who seem to have forgotten that they were once on the other side of the street. They have forgotten what it was like on the way up the ladder. At the same time, managers should make it clear to the person being reviewed that it is not economically wise to pay top dollar to a person with limited experience. This can be a tough issue for a manager to present to an employee, but it goes with the job. A manager must do what is correct and fair, not what is popular.

POSTING POSITIONS

One way of removing the cloud that hangs over salary compensation is to post the employee position status chart, Figure 8-1. A common topic of conversation among employees is to discuss the value of other individuals and what they "think" these other employees are earning. One of the first things I did when I became a manager was to request the salary status of each employee for which I was responsible. At the same time, I placed the individuals into groups based on their position description/capabilities in the company.

Position	Personnel	Status	Salary Range
Service	Al T.	X	$45,000
Installation	Unassigned	O	to
Engineering	Bill	X	$55,000
Estimating	George M.	X	
Senior Engineers	Jay	X	$40,000
	Steve	X	to
	Sylvie	X	$50,000
	Unassigned	O	
Engineers	Matt B.	X	
	George B.	X	$30,000
	Al A.	X	to
	Unassigned	O	$40,000
	Unassigned	O	
Engineers-in-Training	Andy	X	$24,000
	Unassigned	O	to
			$28,000
CAD Operators	Mike	X	$20,000
	Pete	X	to
			$30,000

Figure 8-1. Example of an employee position status chart.

This organized method of posting the employee's name, position, and salary range provided me with some interesting and unexpected results. After compiling this information into the position status chart, I spent time studying the results. From this review, I was able to develop a course of action, which would provide consistency to the internal employee organization. One of my goals was to place people into categories based on the job description and/or expectations of the company.

When I was ready to post my results, I had already reviewed my findings with my boss and had the support of management to follow through with my plan of action. By posting this chart and explaining the intent to everyone in the group, I was able to stop the gossip that employees were having regarding salaries and capabilities. If I was wrong, I was wrong on paper, and this provided the affected individual the opportunity to meet with me and discuss his or her status and/or abilities. When I posted the position status chart, I let everyone know that the listing was arranged in order of ability. In addition, I let everyone know that salaries would be arranged so that the people within a specific group would all be within ±10% of their fellow employees in that job description. This chart eliminated much of the internal gossip and generated an opportunity to initiate communication between management and the individual workers.

Another goal of mine was to use this position status chart as a tool to ensure fairness to the employee and employer. The position listing clearly highlighted that we had some discrepancies in our salary structure. The most noticeable discrepancy, from an employee's point of view, was the salary of one of the engineers-in-training. This woman had joined the firm right out of college and had applied for a position as a draftsperson, because the company had advertised for that need. Within a short period of time the company was using her as a design engineer. She never made an issue of her salary, because she recognized that jobs were not easy to find at that time and she was learning on the job. But as a manager, it is important to look out for the employee as well as the employer. By using the position status chart, I was able to show that we were not paying her the correct amount for the job she was performing. In fairness to the employer, the company did not intentionally underpay this person. Instead, the person was underpaid because she had applied for a lesser position and eventually was given the opportunities to educationally advance within the firm. She was a graduate engineer, right out of college, but the company wasn't looking to hire an engineer-in-training at that time. We made a substantial change in her salary almost without question by upper management. I was able to do this because of the position status chart. It was obvious that an error had been made over a period of time, and the company was prepared to correct it.

This consulting firm, like many non-union firms, often lacked the organizational consistency to monitor the performance and pay of individuals. Hvac/r design, service, and build companies that use union help usually won't have this problem until they start making special arrangements to pay selected members above the union rates. Management should not focus on the employee who yells the loudest. Instead, managers should be pro-active towards all the employees. Often, a person will come to work, do his or her job, and expect to be treated fairly. Sometimes this type of individual can be naive by assuming the employer is organized and observant of each and every worker. Unfortunately, management is given more credit than it deserves. Seldom does it interpret the quiet person as being unhappy with his or her job and/or salary. Interestingly, I have found most "quiet" employees are usually women. As women continue to pursue more responsible positions in business, I hope and expect these employees will be more aggressive in establishing their fair value salary.

Position posting is not new to the larger firms. When they are looking to fill a position, the personnel manager will post the opening. The worker who wants to advance can initiate a discussion with management as to what is needed to move forward. To other employees, the listing serves as a reinforcement that they are being recognized for the jobs they perform. To the manager, the list is a road map to the financial structure of the staff. Smaller companies also need to develop methods for posting the employee position status chart, because the same advantages will pertain to smaller companies as they do to larger companies.

THE FINAL WORD ON SALARY INCREASES

It has always been my position when going into a review that a manager should never change an employee's salary increase because he or she threatens to quit. I believe in doing your homework when preparing for an individual's performance review. By using the position status chart and walking around the office communicating with the employees, a manager will be well prepared to review a worker's performance. Another important item good managers should know is what the market is paying employees. This can be difficult to achieve but not impossible. By networking with other employers, prospective new employees, and in-house employees, a manager can formulate a salary structure with reasonable accuracy. This format should be kept up-to-date through regular discussions with other people in the business.

Once you have your salary structure in place, employee salary increases become easier to determine. By putting it down on paper, you will find your recommendations to be more readily acceptable by upper management. I always made a point of meeting with my boss to review my salary increase recommendations for my workers, and we would come to an agreement before I met with the employee. This upper management meeting served a number of purposes. First, it is always good to get a second opinion. Second, what we agreed upon is what we went with. There was no going back after the review/salary increase was given to an employee. If the person wasn't satisfied with the increase and eventually gave notice of quitting, we would let them go. I don't believe an employee should be offered more money to stay just because the person submitted his or her resignation. All salary increases should be made at the time of the person's annual review. Make your best offer to that person based on his or her position description and past year's performance.

This approach worked well enough for me to convince upper management of my recommendation. At the same time, if my boss disagreed, I would move forward with the review salary increase based on whatever value was agreed upon by my boss and myself. Even if I didn't agree, I supported the company position. This is important in management, because I have often heard other managers tell the employee being reviewed that it was **their** boss's fault. Workers just don't believe this excuse. If the manager truly believes the person should receive a larger salary increase, then the manager should have succeeded. Only during a major recession did I run into a situation that I found difficult to support. At that time, most employees recognized how fortunate they were to have a job in the construction industry. As the economy began to turn around, I pursued increases for a select few whom I considered to be underpaid. I wasn't as successful as I hoped, and I made it clear to those workers that I would continue to pursue salary increases but that it would be at their next annual review.

You need to recognize the position of management even if you don't agree with it. At the same time, you need to be prepared to lose those employees if a better opportunity arises for them. The company must stick to its position on the salaries and not divert from this position. Giving more money to someone who has given notice should never be acceptable to you as the manager.

This process of coordinating and agreeing with upper management on the amount of money a person should receive doesn't mean you don't listen to the employee during the review. Do your homework long before the review meeting. Meet with the boss and use that advice as a good sounding board for your meeting with the employee. You should enter into an employee review prepared and of a mind set that what the firm is offering is fair and reasonable. I have given over 400 annual reviews, and only twice have I had the employee convince me to reconsider. In both cases, I listened and took note of the issues the person brought out. In both cases I went back to the boss, discussed the issues, and we agreed to increase the person's salary to the level the employee had requested.

Finally, it is wrong for a manager to make a decision based on someone's personality and to not evaluate the person, position, and performance thoroughly. It is also wrong for the manager to be concerned with offending someone. As the person in charge, it is important that you do the best job possible the first time. At the same time, if the employee is not satisfied with the review, even though adequate time was spent discussing his or her performance and value to the company, the person may soon resign. Often, a company will attempt to talk the person out of leaving. Talking can never hurt, but making any other concession is a mistake. Management's credibility will suffer, and it is very possible that the person will still leave within the year.

FAIRNESS TO THE INDIVIDUAL

When I took over the engineering group, we had one individual who carried a gun, which was something I didn't see a need for in an office environment. Having a person whom I considered to be so very wrong was a true test of my ability to treat each employee equally. This individual was not someone whom I would have hired, but in taking over the group, I inherited him. When working with him, I always focused on the work at hand, how he was doing in his job, and his job-related goals. When his review came up, I truly believe I used the same methods to evaluate his performance as I did with the other employees. Within two years, he left the company for more money than what I believed his position should pay. This occurrence was not unusual, as other companies would look to our firm for talented workers. Sometimes they were successful in stealing the person away. In this case, they got an employee with limited growth potential. When he gave his notice to leave, it was one of the few times I was truly happy to see someone leave the company. I always considered this person as my ultimate test and success of being a fair manager.

Another example of fairness is to emphasize the need for the individual to do the job and then get paid for it. Companies often pay top dollar for people before they have proven themselves. Compensating a worker beyond the salary range, as indicated by the position status chart, is wrong. Paying a person for achievements beyond his or her own proven track record of performance will disrupt your management plan. Every once in a while, a person will want to take on a more responsible role in the company. At the same time, the employee will ask for or expect to receive a salary commensurable for this service. If you sit down with the individual and outline a plan of action and a time frame to meet the goals, then you can also agree on a financial goal. I don't believe anyone should receive an "advance" based on the excitement of the new position. It's easy to give someone a salary increase, but it's very difficult to take it back. The damage of reducing someone's pay is never forgotten.

Employees are often skeptical of management and its "ulterior" motives. Credibility is essential in management and is needed at all times. Saying one thing and doing another is always a problem that the workers will make note of and spend time discussing inside and outside the office. Being fair to each and every employee is like walking a fine line. As the manager it is always your decision, but it never hurts to ask someone else for theirs. This is particularly important if you are going to be fair to yourself. You won't always make the correct decisions, but if you are indecisive, then it's time to get a second opinion. If you maintain a 51% success ratio with the people you are interviewing, then that is about as good as it is going to get.

Don't Worry About Losing Employees

I try not to worry about losing a person to another company. If you believe you have done everything correctly, then you need to believe your employees will do what is right for them. When a review has been completed, a manager should have a sense of where he or she stands with the individual who was just reviewed. If it went badly and the employee isn't satisfied with his or her salary, but you believe the worker has been treated fairly, there is little more you can do. Paying employees more than they should be receiving for their job description is a major error on the part of management. I have always said that I would rather see someone leave the company, get another job, and receive more money than he or she is worth, than to give in to the employee's request. So often over the years, I have seen people leave a company and six months to a year later hear that they have been laid off. I believe that if the individual can get more money elsewhere (in a good economy they usually can), then let them go. It is now someone else's problem that the new employee is overpaid. When the time comes to let someone go, it is now another manager's problem.

As the manager, you are not doing anyone a favor by overpaying them. All businesses have their ups and downs, and when business is down, the overpaid person is usually the first place you look to cut company costs. I don't enjoy the responsibility of laying off employees when business requires it be done, and I certainly don't enjoy laying off someone with a high salary that I helped elevate! The reason I feel so strongly on this issue is because the person who is laid off is deeply hurt. Managers don't need to put themselves in this position.

Within a month of becoming a manager, I had to lay off a number of people. I naturally inherited a few of these "high priced" people, some of whom I thought were overpaid and some I believed to be underpaid. Those who were underpaid didn't have to worry about being laid off. This is probably the only time an employee will breathe a sigh of relief that they are not getting their due share. The problem I was confronted with was the "high priced" employees who had to be laid off. The first place to look when confronted with this issue is the position status chart. This organizational chart will provide you with a measure of the individual's performance as it relates to salary. In my example, we clearly had a number of people who made an excessive amount of money for the jobs they performed. But, it was still not an enviable task to have to let these people go. Each of these individuals was a very nice person, had outside responsibilities (i.e., mortgages, car payments, etc.), and considered themselves valuable to the company. I think it was this first cut back experience that helped me develop my position of letting dissatisfied employees go somewhere else to be overpaid, so someone else would have to lay them off. Giving out raises is easy. Not giving out a raise goes with the management territory!

A manager should always be concerned with employee turnover. Frequent changes of personnel probably means the management is doing something wrong. Frequent changes in personnel will also affect the quality of the product, engineering documents, timely installation periods, and/or service response. Turnover affects

quality control, because each new employee will need a period of time to become familiar with the standards of doing business at that specific company. At one construction company, the firm lost eight of their experienced project managers, which was approximately 75% of the construction management division. These people were very familiar with the company standards, and each had more than ten years of experience at this firm. Losing the nucleus of the staff takes years from which to recover, and in this example, the construction firm has yet to make it back to the lofty position it held in the '80s.

Losing staff cannot be taken lightly. However, if managers do their homework, they will be in a better position to maintain a staff. At the same time, the manager cannot make concessions to workers because they threaten to leave if their requests aren't met. The annual employee review meeting and managing by walking around are examples of how a manager can perform his or her job of satisfying the employees' needs. If you are fair to employees, provide professional development direction, and pay employees accordingly, then it is up to them to reciprocate by remaining at the company. If this isn't good enough, then it is best if they move on, because there are plenty of other people looking to get ahead.

SUMMARY

Managers must set the example by meeting their own deadlines. From an employee's point of view, the worker's annual review is one of the most important manager responsibilities to be met. The majority of employees in the hvac/r industry are inherently driven by job satisfaction. Draftspeople take pride in the details, design engineers eventually want to be engineers, service technicians solve unscheduled shutdowns, maintenance technicians want systems to operate efficiently, and project managers strive to meet project completions on time and within budget. All of these job descriptions and responsibilities can be fulfilled by the employees. The one significant responsibility that these hvac/r personnel want from their managers is to receive, in a timely manner, feedback on their performance and salary increases in accordance with their job function.

The review process is an excellent opportunity for management to meet one-on-one with each employee annually. Managers should maximize the time they have during each review to maintain the corporate foundation, i.e., long-term employees. At the same time, the manager should recognize that change can be good. When a person chooses to move on, it is usually in the best interest of the company. If, as a manager, you have done your best to give direction to and be fair with each employee, then you've done all that can be expected of you.

Making exceptions to the rules will usually come back to haunt most managers. However, on two occasions I can recall changing an employee's salary increase, because some valid points had been brought out during the review which I had not considered. The manager is not always right, but if you have done your job, management will be respected for the role it performs. If you have listened to your employees and taken note of their performances; if you have been fair and honest with yourself and your workers; and if you are able to sit down and discuss both the good and the bad with each person, then the annual reviews will be annual successes. Sometimes a review can even be fun. I remember one particular employee who pulled out a picture of his young daughter before we got started with his review. Carefully placing the picture in front of me, he pointed out the "hungry" look on the little girl's face and the fact that she wasn't wearing shoes. I assume she got new shoes after he had his review!

NOTES

[1] Blanchard, Kenneth, Ph.D. and Spencer Johnson, M.D., *The One Minute Manager*, William Morrow and Co., Inc., New York, 1982.

Chapter 9
Recruiting and Laying Off Employees

Recruiting the right people is the most difficult management task you will be confronted with when managing a company or a group. In all my years as a manager, I can recall vividly only four individuals that I should not have hired. Over the years, there have been some good selections and some great selections, but for some reason, I always seem to remember my worst selections. The problem with recruiting people is that you really don't know how good they are. True, you do have the person's resume and references. However, these two sources of information give you only an introduction into the person's past performance. What you don't know is how good they are for the position you want them to fill. Like recruiting, letting people go is a task that goes with the job of being a manager. The information in this chapter will offer you some insight into the recruiting process, as well as when to let a person go.

THE INTERVIEW

When you become a manager, one of the things you need to do is prepare yourself for the hiring of new employees. The job description checklist mentioned in Chapter 6 is an excellent tool for you or a personnel manager to use when interviewing a prospective candidate. With each job description, there should be a list of tasks necessary to perform the job. Developing these checklists gives you the opportunity to think through each position's responsibilities. A by-product of developing these checklists is that you will have put down in writing exactly what you expect from each person and each job function. Immediately, you have a means by which to measure the employee's performance. Because this is a standardized process, a manager can use this same "script" as a tool to reprimand an individual for nonperformance.

At one company, I had the benefit of a personnel manager, better known today as a human resource manager. When looking to hire a person, the personnel manager always performed the first interview. This screening process often saved me the time of meeting with the candidate. If the other manager believed this person met the criteria advertised, I would then meet with the person. Together, we would both interview the potential employee. This process provided a couple of benefits for the company. First, the personnel manager became more experienced with the qualifications being asked of the candidates, which helped this particular manager understand the hvac/r needs within the firm. Second, I had the input of the personnel manager who was usually listening while I asked questions during the interview. This was good feedback for me to here what this manager had to say about the interview. Also, when it came to company policies and benefits, the personnel manager was more familiar with the details associated with the firm.

If the personnel manager and I were interested in making an offer to the candidate being interviewed, I would say to the other manager, "I think this individual is very similar to Matt B." Referring back to the position status chart in Chapter 8, my reference to Matt B. (who was already working for the firm) was directed toward his position as an engineer, the salary range of $30,000 to $40,000, and more specifically, the fact that Matt B. was receiving $35,000 annually. If the candidate was applying for a position as an engineer, this cryptic message told the personnel manager that the candidate should be offered a salary of $35,000 ±10%. If the person was applying for a more responsible position, the message meant that the candidate was not qualified for the job. If the person was applying for a lesser position, then the message meant that the individual was over-qualified for the job.

Another benefit of using the position status chart was that it allowed us the ability to control the salaries of the people being hired. There was a consistency with the existing employee salaries and the salaries of new hires. If the candidate accepted our offer, he or she would come into the firm aligned with other equals. On a few occasions when interviewing prospective employees, the salary compensation the candidate was looking to receive was significantly below my salary structure. With one particular individual, the personnel manager and I told him he was asking for much less money than what the position paid. We indicated that if he accepted our offer at the salary he was asking for, we would revisit this issue in three months and determine if he should be earning a salary more in line with our salary range. He said he needed to think about our offer. This answer surprised me, because we were telling him that in three months we would give him more money than he was asking for based on his performance. He called us within a week and told us he was going to take a job at another firm. What I learned from a manager at the other firm was that this candidate took the information we had provided him and went to his next interview asking for more money than what we agreed to pay him after three months. The other firm paid him the higher salary!

In regards to the interview, job description checklist, and position status chart, the personnel manager and I would always ask candidates what their salary requirements were, but we never made our decision based on what they said. Our salary ranges were set, and the exact offer was based on how a candidate matched up to existing employees with similar capabilities. The reason we asked what they wanted for a salary was to receive feedback on how we matched up with other company salaries. Unless the staff is union, there is no set salary compensation for workers in the hvac/r industry. The best you can hope for is to get within a range that is adequate for the job description and appropriate for the economy in your region.

WOMEN IN THE WORKFORCE

Having had four painful experiences with hiring the wrong person for the job, I am always amazed at how some managers can be so selective in their hiring process. As a manager, I have hired four people who absolutely did not work out with the company; all of these people were men. Each of them failed to live up to what they said they could do. At the same time, I have had a few women work in my group, and each of them proved to be very good at what they did. More importantly, they each possessed the ability to continue to grow in the business.

As a manager, it is your responsibility to hire the very best person for the job. Because the process of hiring is so difficult, burdening the job criteria with superficial elements will compromise your hiring process. You are held accountable for the performance of the group you manage, and limiting your selection of potential candidates limits you as a manager. It is your responsibility to be the best at what you do and to be a leader to your employees, so they can be the best at what they do. The result of putting forward your best effort is an excellent product. Discriminating will only compromise this product.

An interesting question that was presented to me prior to hiring a female engineer was "what if she becomes pregnant?" As a manager, the best you should hope for is a person staying with the company five years. If you are always successful with keeping all your best employees for this period of time, then maybe this concern might enter your mind. Based on my experience, you will be hard pressed to keep **all** your best employees for five years. The hvac/r industry is a small community of workers, and the word gets out as to who is very good and who is looking to hire a particular specialist. Even when you think you have done all the right things to keep an employee loyal to your firm, people are going to move on. Worrying about losing a worker to pregnancy will be the least of your worries. The better your staff is, the more chance you have of losing someone. You need to focus on keeping all your best workers, men or women! Other companies are looking for the best people too.

I have lost only one engineer for approximately twelve weeks to maternity leave. I hate to think how many really good employees I have lost to the competition! A manager can't put limits on who should or shouldn't be hired. Aside from the fact that it is illegal and unethical, you need to have the best people, and the best people are not all men. In fact, some of your worst performers are going to be these same men. When hiring a person, use the specific job description checklist and the position status chart. These two manager's tools will clearly identify the experience necessary and appropriate salary for the vacant position. If you are interested in hiring a trainee, look at the person's grades and his or her participation in outside activities. In addition, look to see if there is any indication as to how the person's education was financed. I like to see that students have paid for a portion of their own education and didn't have it all paid for by someone else. These factors have far more weight and value than whether the individual is male, female, white, black, etc.

Another concern in regards to women is whether they are strong enough or athletic enough for the job. This comment is usually made by some man who is 40 pounds overweight and short of breath. I am always amazed at the physical strength comments directed at women by this type of individual. A manager must inform each prospective employee of the physical requirements of the job. The position may require carrying heavy boxes of inventory, climbing a ladder, rigging a fan into place, or working on staging. Anyone with some strength should be able to handle most jobs. Heavy items should be rigged into place using the correct tools and a sufficient number of laborers, maybe a crew of women workers. No one should be expected to lift too much weight themselves, because doing so costs companies thousands of dollars a year in workman's compensation insurance, lost production time, and possibly the loss of an employee to a permanent disability.

When you hear comments regarding a woman not being able to physically keep up with the work, consider the source. There are very few individuals in their thirties who are in good physical condition. Whether you are a manager of a construction crew or a warehouse staff, you should be leery of anyone who may be a physical liability to the operation. A person who does not follow company safety rules, such as using safety glasses, back brace supports (when lifting heavy objects), gloves, etc., should be given an initial warning. After that, more stringent measures need to be taken. With the proper gear, a trained individual can always perform the work. The bottom line is **a woman can perform any job**

in the hvac/r industry. All one needs is the desire to do the work, the training to succeed, and the opportunity to perform.

Another stigma for women in the workforce is the issue of day care. Although child rearing is a joint venture by men and women, it usually seems to become a woman's problem. In the hvac/r industry, many firms do not offer day care at the workplace. This is a topic that has been around for many years, but only recently has it truly become a company issue. I have yet to work in a firm that has offered day care to the children of employees. Certainly, there will be an added cost to accommodate this benefit within the company. Day care personnel, space, and supplies will impact the corporate overhead costs. However, the personal comfort of knowing that you have this in-house benefit to assist you while performing your job must certainly bring value to the company's effectiveness. Employees' minds are at ease when they know their children are under the same roof being well cared for.

One reason day care has made its way to the forefront is directly related to the number of women working and the number of women having more responsible job roles than in the past. Women are slowly infiltrating the management ranks, and with this infiltration comes the need to allow them to commit the time necessary to fulfill their job requirements. Being in a more responsible role provides them with the added strength to request or require the company to accommodate their child care needs. A firm is continually looking to hire the best workers available. If that means making some concessions to get the maximum hourly output from these high performers, then day care should be a major consideration.

Working mothers are usually striving to achieve job satisfaction while meeting parental obligations. This issue adds to the need for a manager to not only be sensitive to the issue of day care, but also accommodate that responsibility. Recognize that this is a business issue that must be resolved. As the manager, you should be pro-active towards resolving the problem. I know of one business owner who would rather have his secretary bring in her three-year old daughter when she cannot take the child to day care than to have her stay at home and care for the child. When this problem occurs, this owner has two choices, to have a secretary or to not have a secretary. His decision was to let the child come to work with the mother. After all, the little girl can't possibly make more noise than an engineer having a bad day!

WHEN YOU HAVE HIRED THE WRONG PERSON

If **you** have hired the wrong person, then it is your responsibility to correct the problem. However, the odds are in a manager's favor if he or she initially uses the job description checklist, which lists the job experience and requirements for each position. It is during the interview that an unqualified person may give conflicting answers to the checklist questionnaire. Corroborating the person's references is also very important. In my earlier years, I was naive enough to think that people didn't lie on their resumes. For example, the resume of one candidate indicated that he was an out-of-state, professional engineer with a mechanical and electrical degree and also a part-time professor. He turned out to be nothing like his resume. In fact, the registration number belonged to a person who had died years earlier. The more responsible the position, the more important it is to verify resume accuracy and references.

Once you have hired a person, maintain a close watch on the individual's work performance and work ethics. It's a lot easier to correct a hiring error early than to allow someone to remain on staff for more than six months. Also, ask for feedback from the person's co-workers. Do they enjoy working with the person? Does the person have potential? Is the new employee very knowledge-able? Your interest should be presented in a positive manner. After all, you do want the individual to succeed. Presented in the wrong context, your interest may send an incorrect signal to the other workers and eventually back to the new employee. At the same time, offer feedback to the new worker so they know you are interested in what they are working on and how well they are doing.

If a person is not working out, it is important that you meet privately with the individual to discuss your disappointments. Be very specific with the person about issues that have resulted in your meeting. With trainees, the issue at-hand usually is their "slow" progress, and most of these problems can be corrected over time. Remember to reiterate the following points:

- You want to get the most out of them for their benefit, as well as the company's benefit.
- They are going to be more valuable to themselves, as well as to the company, in one year and in three years.
- If they don't get ahead, it will be their own fault, not the company's, because they have been given the opportunity to succeed.

I have used these three statements continuously through the years. I have used them when encouraging a person, reprimanding a person, and during annual reviews. Each statement reflects a positive goal that you want the employee to achieve. When workers aren't performing, they must be reminded of these mutual goals. Then they have been warned.

If the employee's performance continues to be poor, it is important that the manager take corrective action. There are two alternatives that can be implemented. The first alternative applies to a person who came into a responsible position with limited experience, such as a graduate engineer with a few years of experience. Their error may be that they applied for a job significantly beyond their present skills but somehow convinced the manager during the interview that they could do the work. In this example, you may consider keeping the person but reducing his or her pay. Another compromise may be to indicate that the person will not be receiving a salary increase after the first year of employment. Both options should be presented in a manner that reflects your willingness to continue to work with the person and to help the person succeed. At the same time, it is the manager's responsibility to be cost effective with the salaries of those in the group. I have had to implement similar corrective measures in the past. In most cases, the worker continued to be employed and worked to improve. In a couple of cases, the employees chose to look elsewhere for employment. When employees choose to look to other companies, chances are they aren't going to tell you, and you won't know until they give their notice. This should only be a slight inconvenience to your operation, because they haven't been at the company long enough to create a workload problem. In addition, because these people were not working out, you may have already assigned someone to assist on the projects that these newer employees had been working on.

The other option is to terminate employment. When a manager is in this position, termination usually is the correct course of action. In most cases, workers come into positions with a significant amount of experience and the qualifications to perform the work. Likewise, if a person isn't working out, it is correct to assume that this person does not have the ability to improve in the near future. As a result, they are not good candidates in which to invest company time and money. The decision to terminate is always difficult and should be made without prejudice. However, after it has been done, you will know you made the correct business decision.

NETWORKING AND THE RESUME

Networking and resumes are two of the most important sources of information when it comes to hiring a person. A manager always should be looking for the best workers. Since no one company has cornered the market on having the best people, you need to stay in touch with the marketplace. Through a network of key individuals, a manager can stay abreast of the talent associated with his or her business. In addition, managers should solicit resumes on a random basis, even if they are not specifically trying to fill a position. Whether the hvac/r economy is prospering or in a slump, the opportunity to hire a very good worker will frequently strengthen your company's talent base. However, when the marketplace is slumping, a manager may be able to attract a talented person, because they see that your firm is strategically planning while others flounder.

Networking is an excellent first step in finding the person to fill your company needs. Whether you are in service, construction, consulting, maintenance, or manufacturing, the business community is amazingly small. For this reason, experienced managers seldom use personnel recruiting firms. Instead, networking within your work area can be an invaluable source of personnel information. For example, sales engineers who call on several hvac/r firms can be valuable sources of information. People who have just joined your firm are also a valuable source of information. At the same time, when looking for a project manager, a consulting engineer may have the name of an individual they have worked with recently on another project with another construction firm. The construction company may return the favor by dropping a name or two of experienced engineers that they are familiar with at other sites.

Always be interested in who is out there. The better your staff, the easier your job is going to be. The more experienced you become, the more you will appreciate having very good personnel. Listen to what other people have to say about individuals in your line of business. People outside of the company that you respect for their hvac/r talent can offer some excellent observations of other employees. A manager can be assured that these same people have a mutual interest in seeing system designs, project installations, and operation and maintenance performed efficiently. Everyone can appreciate a talented person, and they don't hesitate to speak up and compliment this person. As a manager, you should look to those talent sources for your information. I recall one individual who took pride in knowing who was considered very good and if they were "movable." He was one of the first people I called when looking to hire someone.

When looking at resumes, managers have in their hands another valuable source of information. If you believe in the concept of keeping an employee for five years, you should also look at resumes in a slightly different light. After reading a person's educational background, look at the number of jobs a candidate has had during his or her working career. If they show a change of job every two to three years, I immediately discard the resume. You have to say to yourself "what makes me think I can keep this person any longer than the previous employers?" The fact is you won't be able to, so don't even waste your time. I want a person who has demonstrated a willingness to stay at a firm for a sufficient time period and seen a project through from beginning to end. Often, a person will join a company, be assigned a specific task or project, and work on that project until a better job offer comes along, at which time they leave to "grab the opportunity." Job jumpers, as I like to call them, are more common during a good economy and usually don't leave companies when business is slow. Job security at the new firm may be questionable, so these people will tend to stay where they are.

If you see this type of pattern on a resume, chances are the person never stayed long enough at one firm to learn any task from beginning to end. These people may reflect a vast and diverse array of experience, but their resumes document that they never put in sufficient time to be truly skilled at what they say they can do. A person needs to learn while on-the-job, so he or she can improve on past educational experience.

The hvac/r industry requires a combination of formal schooling, including vocational technical schools, trade schools, colleges and universities, various technical courses specifically related to the work being performed, and on-the-job training. When hiring so-called experience, a manager needs to be conscious of the time element necessary to become proficient at the various tasks a person claims to be able to perform. You need to be able to recognize and question the resume at the interview. A very good way to question the individual is to use a job description checklist, which was discussed in Chapter 6. By asking candidates very detailed questions, you may find inconsistencies in their answers. At the same time, I find that the checklist is not intimidating to the candidate, because it is perceived as an agenda item in the interview. The questioning can be presented in a manner that does not "terrorize" the individual. That's not the intent of reviewing the person's resume.

Applicants may have held jobs for less than three or four years. Those things can happen. I prefer to see that a candidate has worked at his or her last two jobs for a minimum of five years each. If the person's experience and education meet your criteria, then you can assume that keeping the person for a sufficient number of years is a realistic goal. Even a series of three and four year employments may be satisfactory for your needs. These three or four year periods demonstrate a stability bordering on your guidelines; therefore, the candidates should still be considered for the job. A continuous pattern of staying at a job for less than three years may signal a potential problem person. A manager doesn't need to create problems by hiring this kind of person, no matter how good his or her resume may look.

THE EXISTING EMPLOYEE

This is always a very sensitive subject for a manager, because loyal employees aren't necessarily good employees. While you are trying to keep your very best workers, a company will inherently keep some of its marginal workers too. Managers frequently miss this development within the company, because the person in question has been a good worker for a significant number of years. Over time, this type of person reaches their peak ability and peak earning power for the job position. At some point, they slip into a position of employment comfort and slowly lose the drive to continually improve. This type of employee is more apt to develop within a consulting firm, construction company, or service company. It is less likely to occur on the assembly lines of the manufacturing segment of the hvac/r industry. For assembly line work, you need to have a person with the skills to perform the repetitive tasks associated with the job. Limited growth potential exists for the majority of factory workers. These workers take pride in what they do daily, as opposed to the other segments of the hvac/r workforce who strive to have more client contact and one-on-one dealings with management.

In the consulting, construction, and service businesses, the existing employee is likely to be the person representing the firm. The demand to continually improve client relationships requires the employee to continually improve. If these employees don't continually improve, they begin to take a "backseat" in their client maintenance activities. At the same time, it is not unusual to be able to advance up through the ranks of business. Because this is a normal route within the service industries, a manager expects employees to continually expand their skills. It is this management expectation that can turn to employee disappointment if the person "peaks out" along the way.

When employees become comfortable with their jobs and their performances level off, it usually occurs over a long period of time. This can be a casual transition that occurs as a person grows and sets his or her priorities in life. This kind of transition frequently goes unnoticed by the manager. The person in charge may perceive this shift in priorities as a drop in performance. It is not unusual for a manager to make the mistake of continuing to provide salary increases based on past increases. Instead of recognizing that the person is reaching a "comfort level" and his or her drive to advance has waned, the manager will carry on and expect more than the worker is committed to giving.

These individuals can become a financial liability to the firm over a period of time. Just like the manager who doesn't recognize this "leveling off" performance, the employee doesn't recognize this gradual change and the compromises that go with accepting status quo. This individual is still expecting the same salary increases that he or she received before and the same opportunity to advance. Neither the worker nor the manager recognize the change immediately and a great deal of time can go by before one of the two people recognize that things have changed.

It is very important that a manager be in tune with employee performance, goals, and expectations. From these three topics, a manager can recognize a "leveling off" and discuss this issue with the person. If not caught in time, a disgruntled employee may leave the company for all the wrong reasons. At the same time, a manager may lay off a good employee mistakenly, citing poor performance as the reason.

Lay-Offs Are Planned

This responsibility may be one of the toughest tasks you will have to perform as a manager. Long before lay-offs occur, you should be able to recognize that they are going to happen. A manager needs to be cognizant of this issue and be pro-active when it comes to terminating workers. What I mean by pro-active is to be on top of the issue early, set the strategy, prepare alternatives, and get the job done.

A manager should be planning the work schedule in time increments appropriate for the business. By using the staff and workload requirements schedule discussed in Chapter 7, the person in charge will be able to determine if a lay-off is inevitable. To minimize this dilemma, management should consider the strategy of being slightly understaffed at all times, approving an overtime work schedule to resolve the "peak" workload when necessary. For example, consider four, ten-hour days, eight hours on Friday, and a half day on Saturday. The next measure is to have a pool of subcontractors available to come in and assist with the excessive or prolonged peak workload periods. At some point, you should also consider hiring an additional worker when you believe it is a viable choice. Likewise, when companies experience "valleys" in the workload, using the same staff and workload requirements schedule can help managers create a strategy to work through this lull in activities. Encouraging employees to take their vacation time and changing to a four-day work week are two options.

A slowdown in work is also an opportunity to reduce the work staff by terminating those individuals who don't have a future in the company. Employees who have become comfortable with the company, take advantage of the firm by coming in late and leaving early, and constantly complain about the company are good candidates to lay off. A manager wants personnel who pose questions and provide solutions; who contribute to the effort and do not detract from it; and who recognize that work is a responsibility, not a benefit. Less than a month after I was promoted to manager, the company was facing a lay-off. I was not as prepared to handle the situation as I would be in the future, but it was my responsibility to terminate a number of employees because of a work shortage. I recall one individual who was terminated who should have been let go long before I became a manager. This person fit all of the requirements for termination, and the slowdown was an opportune time to lay off this person. When he left, a number of the employees where glad that the company had addressed this particular issue, because he was a distraction to those who came to work and enjoyed their jobs.

When it comes to laying off a person due to a work shortage, try to find the individual another job somewhere else. As the manager, you will know that the lay-off is coming, so call around to a select few businesses that you know are busy and where the person in charge is someone you can confide in. The last thing you want is to have your employees hearing about your lay-off from someone outside your firm. Remember, the hvac/r industry is a very small community! Your attempt to place the laid-off worker in another company will be greatly appreciated by that individual. Obviously, this person would have to be someone whom you would recommend when calling around to other companies. If this worker is being terminated for other reasons, along with the work shortage scenario, then the manager should not be trying to help the person get another job somewhere else.

BE FAIR

When faced with the need to downsize the group, a manager should always be fair and never base his or her decision on a prejudiced point of view. Remember, you need the best help available to succeed at your job. Don't be foolish enough to discriminate against individuals because of their race, sex, or religion. At the same time, don't let a very good worker go in order to keep a loyal but less productive worker. The need to reduce your staff is always a difficult situation. A manager who complicates this situation by letting unrelated issues influence his or her decision is only making matters worse. Don't make the problem any bigger than it has to be. The goal is to implement this change to improve the company's financial performance. If you give away a major leaguer and keep a minor leaguer, the next person to go may be you, the coach!

Fairness also extends into the area of manager-employee relationships. When faced with the lay-off of a person you don't personally care for, a manager must focus on the business issues at hand and put personalities aside. Again, you need the best help available! Not getting along with the boss doesn't necessarily mean that you let a person go. I have had a few employees, thankfully not many, with whom I did not share the same point of view. You must continually focus on making the correct business decision. You don't have to go home and have supper with this employee. You simply have to work amicably together. If the person is someone who others don't get along with but is proficient at his work, then he's not the person you should be looking to let go. Instead, it is the non-performer, the overpaid worker, and/or the person with the least potential to grow within the firm who should be your first candidates for termination.

In fairness to the company, when another reduction is needed and you are down to a nucleus of employees who are very good at what they do, a manager must look at the employees' salaries. For example, if you have two very good service foremen, one with twelve years of experience and the other with seven years of experience, the question of salary will come into play. If they both do the same type of job and both have sufficient experience to be accountable for that job, then who makes the most money? You may find that the twelve-year person is at $64,000 annually, and you consider this to be the peak earning level for that position. However, the seven-year person is making $54,000 annually. If you take into account that neither person appears to have the potential and/or desire to advance to the next level, then you need to keep the seven-year person and let the twelve-year person go. However, you need to be careful when using this approach if the twelve-year person is approaching 40 years of age or older. Your decision to let this individual go must be unbiased in regards to age. This decision is also not based on performance, because both employees are very good at the same job; instead, it comes down to money.

The money decision is twofold. The first issue is that you can reduce the direct labor cost to the firm by $64,000 instead of $54,000. That's an additional $10,000 savings to the firm. The second issue is that the person you choose to keep has additional growth potential, and the person being terminated has reached his approximate peak earning potential, with no further position growth anticipated. If you choose to stay with the higher-paid service foreman, you may have created another problem for yourself a year or two later, because this foreman has nowhere to go financially in the future. If you limit the increases in the next year or two, the employee may become disenchanted with the firm and leave. Worse yet, the foreman may become disenchanted with the firm and stay! Either way, you lose. By keeping the lower-paid foreman, you have financial growth on your side. Realizing that both individuals were doing a very good job and the lay-off is to cut costs, you can maintain a person who has room to grow financially within the firm. Through the process of training the employees, you should have another foreman coming along in a few years to eventually replace the vacancy while controlling operational costs.

LAY-OFFS AND TERMINATIONS

I have never worked for a company that had hundreds of employees. Instead, I have worked in smaller companies where I was responsible for up to approximately 40 employees. Therefore, I cannot offer any alternative to laying off a large number of people other than to do it all at once. Fortunately, most hvac/r businesses are relatively small, and the larger ones are departmentalized into smaller groups. It is in this area that a manager can create a strategy as to how they want to implement the lay-offs.

During the mid-'70s and late '80s, it was not unusual for companies to downsize. In fact, it was not unusual to have more than one lay-off in a year. During those two periods the national economy was very bad. Many hvac/r firms, particularly in the late '80s, went out of business. There were companies that I was familiar with that closed their doors, some after more than a century of business. These have been tough times for many. As a manager, you will probably experience at least one of these terrible periods in your career. When you do, a manager needs to be extra sensitive to the termination process.

In the consulting side of the hvac/r industry, most people are let go on a Friday. The thinking has been that the individual will have the opportunity to look for a job in the help wanted ads of any major newspaper the Sunday immediately following the lay-off.

This particular day traditionally receives the largest number of requests for advertising employment opportunities. As a result, this edition will offer the most job openings. When I worked with a construction firm, the tradition was to have people leave on a Monday, because they could then go to the union hall that day and sign in. The construction crews thought it was insensitive to lay off people on a Friday and ruin their weekend. I never looked at it in that manner, and if you didn't belong to the union, missing the Sunday edition of the help wanted ads could delay your job hunting by a full week.

In a bad economy, don't create a pattern that has employees worrying about lay-offs. At one firm, the workers would jokingly say "you don't want to be called into the boss's office on Friday at 3:00 p.m." The company had made a few reductions in staff and had formed a habit of letting people go at 3:00 p.m. on Friday. This created an uncomfortable environment within the group. Always remember that timing is everything. Hopefully you won't have to reduce your staff more than once in a year, but if you do, be cognizant of the timing.

When it is time to bring in a person to discuss his or her termination, it is best to do so one-on-one. If the personnel manager needs to be involved, let the employee go to that manager after you have met with the individual. You need to be direct and to the point but also sensitive with the presentation of the news. If the termination is because of a lack of work, say so. Possibly you were able to arrange an alternative where the person can apply for a new job. This would be the time to tell the individual of the opportunity.

If the lay-off is because of poor performance and the person is going to be replaced, then the worker is really being terminated. Termination must be dealt with in a completely different manner than laying off a person due to a lack of work. You need to have laid the foundation as to why the individual is being terminated long before it happens. There are numerous legal issues you must be aware of, most of which you can obtain from your firm's human resource group. They will direct you in how to prepare for a termination, including the necessary reprimands and warnings that must be included in the person's personnel file, etc. A manager must be prepared to discuss the issues. In addition, you should be firm as to how you feel about the person's performance (i.e., your disappointment), and make sure the person realizes that this is not one last warning.

SUMMARY

Two of the most difficult responsibilities of a manager are recruiting and laying off employees, both of which require preplanning to be successful. The interview is not necessarily the beginning of the recruiting process. Prior to meeting with any potential candidate, you should have drawn up a job description or position checklist and a position status chart. Both of these management tools will assist you in screening the various individuals applying for the job. Because the candidate may be an "unknown," a manager must be prepared prior to sitting down and discussing the position opening. The better prepared you are, the better chance you have of selecting the best person for the job.

Compare the person's resume with your employment needs and look very carefully at the number of years he or she has worked at other firms. Be very skeptical of people who change jobs every two years. You should set your sights on keeping this person for a minimum of five years. If they have not historically stayed at any one firm for three or more years, who's to say you can change that pattern? In addition, why do you even want to bother trying to change that pattern? You want to pick the very best candidate for the position, so that this new employee will help improve the performance of your group.

When looking to hire an additional worker, don't make exceptions that will compromise your staff structure. The individual must meet the requirements of the job description checklist and the position status chart. You need people who will complement your existing staff within the financial format currently in place. Paying more money to get someone new and forgetting about existing workers can be a common mistake by management in the quest to attract a certain individual. Be fair to **all** the employees, not just the new ones. At the same time, know the marketplace. Keep abreast of what people are being paid and who is very good at what they do.

It is your responsibility to look out for the existing employees, so that they are treated fairly and don't become part of the furniture.

When hiring, prejudice has no place in the interview. It is in your best interest to hire the very best candidate for the job. Once hired, you expect new employees to do the best possible job for their benefit, your benefit, the company's benefit, and most importantly, the client's benefit! Hiring the "wrong" person can be costly to everyone. No one has cornered the market on having all the best people. As the manager, it is your responsibility to strive to achieve that goal. In the process, you will inevitably end up with a blend of people and personalities. If you hire the wrong person, it will reflect on you eventually. After all, who can you blame but yourself?

Always review a resume thoroughly, checking references and verifying the accuracy of the information. Don't take anything for granted. When you have hired a person who doesn't work out, let the person know well in advance before terminating his or her employment. If the position is one of responsibility, a manager has the right to be less tolerant of poor performance. If the new employee is in a less responsible position, patience is required along with constructive criticism and guidance. Remember, you are not paying this worker "top dollar" to learn while on the job. In the management process, don't accept marginal work ethics from "loyal" employees. No one should become so comfortable with their job that their performance becomes mediocre. They have a responsibility to the company no matter what level they are at in the organizational chart. **Stay in tune with each employee's performance.**

Maintaining a staff and workload requirements schedule will allow you to manage the workforce more productively. Be pro-active with this management tool, so that you are prepared to add to or downsize the existing staff when necessary. Don't be unprepared when the need for lay-offs arises. A manager must be in control of the workforce and make business decisions that will be profitable for the company. This juggling act of balancing workload with the number of employees is an integral part of your job. You must be prepared to get through the "peaks" and "valleys" of the hvac/r business efficiently. When it comes time to terminate a person, know the company's legal responsibilities and the individual's rights. Be ready and firm with your decision while being sensitive to the individual. Know the corporate policies and plan ahead.

Chapter 10
Success

I consider success to be a day-to-day proposition. To be successful as a manager, a person must achieve numerous goals and milestones and complete various job responsibilities. In the hvac/r industry, you can succeed with goals, milestones, and responsibilities in many different ways. A truly good manager has a proven performance record that speaks for itself. You may know many others in management who are in the 80% category we talked about in Chapter 2, but the best leaders are the ones whose actions have provided them with their advancements. In order to be one of the best managers, you must get up each day with a positive attitude, and you must possess a desire to lead, a need to do the best you can do, and the aspiration to help see others succeed. You must be able to put forward an honest effort and let your actions speak for you.

GOOD DAYS AND BAD DAYS

If there is one thing I probably have not done enough, it is to stop and smell the roses. Enjoy the moment! If you did something unique, then savor the applause. Because I experienced some low points before I made any significant contributions to the business that I'm in, I have been reluctant to reflect on what I have done well. Managers must continually remind themselves that they are going to have good days and bad days. Enjoy the good days, and learn from your bad days. A secret to management success is understanding that you are going to prosper. Every good manager will. At the same time, you cannot dwell on your achievements too long. It is your responsibility to be the best manager you can be, and that is why you are the person in charge.

For me, I have my good days and bad days. I was fortunate to have engineered at least two first-of-their-kind hvac/r projects: the dual-fan double-duct system and the concept of a building without a heating system. The first project, the dual-fan double-duct system, would eventually be included in the American Society of Heating, Refrigerating and Air Conditioning Engineers (ASHRAE) 1984 *Systems Handbook*. The second project, the building without a heating system, received world-wide attention and even made *Ripley's Believe It Or Not*. Although others continually talk about these two success stories, I frequently reflect back on two less notable jobs. These two other projects taught me more about the business of striving to do the best you can. Fortunately, or unfortunately, I had some sobering experiences with each of these projects that have stayed with me and are a reminder that in the hvac/r business, you can't sit back and dwell on your past achievements. Yesterday you may have appeared to be a genius, but what are you going to do for the hvac/r industry today?

I was approximately 21 years old when I worked on the first of these two less notable projects and was also the lead engineer on a number of other hvac/r designs. I was very busy at the time, but my boss gave me one more project to work on. This new task was a job that was already under construction, but the design engineer had just left the firm. I really didn't want to take on another job, but I wasn't given that choice. At my first site visit, it was brought to my attention that the sheet metal and piping systems were being installed without close field coordination, and a major conflict now existed. I said I would look at the installation and that I would have a resolution the following week.

The next jobsite meeting agenda brought the project architect, the clerk of the works, the client's representative, and the president of the construction company. I presented my solution to the mechanical conflict, drawn on a clarification drawing. My drawing was correct, and it would resolve the dilemma. What I had not anticipated was the reaction of the construction company president. He took my drawing, crumpled it up, and threw it to the floor as he said, "Why wasn't this done correctly, the first time?" Because I was not responsible for the original design, I didn't consider the problem to be my problem! I was embarrassed, to say the least, and shocked at the actions of the company president.

When I left the meeting, the incident kept going through my mind. Fortunately, I was able to focus on the real issue, which was that I had not been prepared to take on this job with a commitment to contribute to its success. Instead, I just went through the motions of getting the job done. As a manager, you will never be fully involved with each and every project in your group. Still, it is your responsibility to oversee its success. If someone in your department fails, you fail too. To blame anyone beyond yourself is bad luck! The success of the group and the project is your responsibility. A friend of mine recently gave me a paperweight shaped like a crumpled-up drawing. It sits on my desk as a reminder of how everything you do is important, and if you are going to do it, do the best you can or don't do it at all!

The second occurrence involved a project where I was guilty until proven innocent. This project entailed designing the hvac/r systems for a major health care facility. At the time, the company I worked for would transfer the responsibility for the project from the design engineer to the field engineer when the project moved into the installation phase. Therefore, during the construction period, the field engineer was the person in charge of the building program. At this new state-of-the-art facility, the customer moved into the building approximately six months ahead of the completion schedule. The Sunday before they opened their doors for business, I received a phone call at home requesting that I come to the hospital. The call came from the administrative vice president at the hospital. After touring the facility, it was obvious to me that it was not ready, but our office had agreed with the contractor that the building was ready for occupancy. Within the first week there were numerous hvac/r problems. Almost immediately, I became deeply involved with the project again and was at odds with the field engineer regarding the dilemma the company was now in. I was 33 years old and sitting at a meeting with the chairman of the board for the hospital, the president of the hospital, and several other notable, elderly businessmen. I sensed that these people lacked confidence in me. All the blame was now being directed toward my design, and I was given two weeks to resolve the problems. It was truly an intimidating moment in my young career.

I sat and listened, understanding the seriousness of the problems. This was a hospital, and many of the occupants were bedridden; therefore, space comfort, temperature and humidity control, air filtration, and space pressurization were essential to the building environment. I listened carefully to the statements being directed at my firm, knowing that this health care facility and its administrative members had tremendous influence in the engineering community. At the same time, I recall a comment the hvac/r contractor made to me at the close of the meeting. He said he would do anything he could to help. I refrained from telling him that he had done enough! The meeting ended with the understanding that we would meet again in two weeks. At that time, all the problems had to be resolved.

In three days, I was able to thoroughly inspect the hvac/r system installation and the automatic controls. I observed the sequences of operation and listed page after page of incomplete and incorrect installations. By the third day, I had made adjustments where I could and directed others to perform corrective work where necessary. At the end of the third day it was clear that the problems were being resolved and that the list of deficiencies fell back onto the hvac/r contractor to fix. He had a team of workers out at the site for two solid weeks rectifying the installation. By the end of that first week, the hospital was satisfied with the progress and our scheduled meeting in two weeks was canceled.

For almost a week, I was guilty until proven innocent. We never reconvened, so I never had the opportunity to receive the recognition for resolving the hvac/r problems. As a manager, you will often be confronted with serious job-related problems. These problems may have long-range negative effects on your company and its business activities. At the same time, these predicaments may not be a result of something that you have done. Instead, they may have been created by people you are responsible for or other members of a project team. You must be prepared to take charge and resolve the issues. To do this, a manager needs to have a wealth of experience to draw upon. You can't lead if you don't have the experience and skills to solve problems. Experience and problem-solving skills come from training, education, and past success.

Just Do It

As stated earlier, a great phrase to manage by is "either do it or don't do it, but don't say you are going to try!" A manager must be out front, leading the troops. You must set the pace. If you don't set an example, then you can't expect anyone else to lead the charge. A manager should be an inspiration to the others. This inspiration should go beyond the average workday. A successful manager needs to keep all the balls in the air. You should be able to juggle your business career with your personal life. What good is success in one area of life if you fail in all the others?

If you possess the attitude that you are going to "do it," you create an attitude and reputation as a person who gets things done. In time, people will say to you "where do you find the time," followed by their comment "I'm always too busy." The difference between this person and you is that you make a commitment and follow through. They will approach a commitment with the attitude of "trying" to succeed. To try implies that they probably will not succeed. **Success is built on a foundation of commitment and follow-through.** Get in the habit of saying exactly what you mean; what you are committed to completing; and what you can or can't do. Building a reputation on positive performance, with the by-product of getting the job done, will bring recognition by others of your skills, attitude, and success.

It is easy to say that you'll "try" to do something, but managers shouldn't mislead people into thinking they will do something they are not truly committed to. Not getting the job done is detrimental to becoming an effective manager. You must be the person to whom your boss, employees, and clients look for results. All this translates into confidence in the manager. In time, your opinions, outlook, and goals can become those of the company and its workers. Just do it!

On the other hand, there will be those tasks that you are not committed to completing. Don't mislead anyone into thinking that you might "get around to doing it." At the same time, there will be job-related responsibilities that you don't care to perform but are still responsible for. In this case, **get it done.** Don't put off those things you don't like doing. Instead, concentrate on the task even if you don't enjoy it or are too busy to get involved. These tasks don't go away by themselves. A manager must continuously strive to accomplish all the assignments within the department. Remember, if you project a negative attitude towards a task, the bad luck syndrome discussed in Chapter 5 may set in. It doesn't take too much of a negative attitude to have the employees following in your footsteps. If you don't want to do something, why would you think someone else cares to perform the work?

The idea of "trying" may sound like good advice, but it's not. Get in the habit of doing or not doing. Encourage the others in your group to follow your advice. Teach them to clearly state that either they can do the job or they can't do the job. Also, remind them that if they don't do it, someone else may be anxious to grasp the opportunity. Don't take it lightly when you say you can't do something. There is a fine line between being negative and non-committal, and a manager can't be negative. Anyone can say they can't do something. This statement sets the stage for these people to avoid failure, because they said "it can't be done." On the other hand, if they succeed, and they usually do, then these people believe they achieved an impossible feat. This is only a charade based on a lack of confidence or a negative attitude. In either case, a manager cannot set an example based on "trying."

Success is based on doing whatever it takes to be the best. A manager must have a positive attitude if he or she is to move into the top 20%. Successful managers aren't afraid to stick their heads above the crowd, and they are not afraid to be wrong. More importantly, a manager can and will make it happen! A by-product of your positive attitude is the commitment of others. They will be willing to work with you, because they believe in your work ethics and your obligation to do the best job possible.

MAXIMIZE YOUR EXPOSURE

When you are striving to do the best job possible, look at the big picture. Be aware of the grand scheme of things. Every day, workers miss the opportunity to raise their heads above the crowd. A successful person will capitalize on his or her performance. To reach the level of manager, you need to make sure that you are getting credit for your high performance. Individuals must subtly position themselves to receive the recognition due them. I consider it a by-product of a job well done if acknowledgment is directed towards you. There is a fine line between putting yourself first and putting the project first. A clever employee can make sure the task is completed successfully and they receive the appropriate credit, while other employees will focus on how they were responsible for getting the job done. Often, it is a matter of how you present yourself. It is very important, when striving to advance within the company, that you are not overlooked for your contributions to the company's achievements.

One of the best ways to receive this recognition is through a published project report or by receiving a project award. The benefits of these methods of awareness can be immeasurable. First and foremost, the company benefits from the time you spend writing an article or gathering the necessary information for an award application. A pro-active company will jump at the opportunity to distinguish itself from all the others. The article or award also represents a third-party endorsement of the firm's skills. This endorsement can be further enhanced by the company's efforts to bring this awareness to its clients. Again, the company wins because the publication or award illustrates to others that this company uses the skills of pro-active employees. If all or some of these benefits occur, then the by-product is your affiliation with the project. If you continue to routinely participate in featuring the company you work for, then you will share in its reputation.

If you don't make the time to participate in these activities, you lose the opportunity to distinguish your company and yourself. I have often encouraged individuals to make the time, reminded them of the opportunity, and then watched the window of opportunity close, because they said they didn't have the time, didn't have the skills, or would "try." Success passed them by, and later they will wonder why you are the manager and they are not. The best managers maximize these opportunities as they become available. This will require additional time on your part, but the job satisfaction is a benefit, particularly if you don't have a boss who believes in praising. You can be assured that others will share in your notoriety and compliment you on your article or award. The advantage of having an article published or receiving an award is that they both "do the talking" for you. You don't have to go around announcing your achievement, others will do it for you.

Another benefit of a published article or an award is that you may be perceived as an authority on the topic. With recognition comes credibility! This may not be an accurate statement, but your achievement denotes an expertise on the subject. This expertise is implied by the endorsement of the agency or firm that credited you with the accomplishment. Others, who have not spent the time pursuing these credits, may not receive the same authoritative status. Managers should be required to aspire to be the expert. It is your obligation to establish yourself as the authority or the spokesperson for the firm. Although very few leaders contribute to this cause, a select few managers will make the time to improve the credibility of the company. You should strive to establish yourself as one of those select few.

It is not required that you have all the answers. As the manager, the more you are recognized as the authority, the more influence you will have to manage the process. This process may be the standards within the company, the quality control, the profitability of the company, or the direction of the company. All these corporate needs are issues to which you can contribute. The more influential you are, the more control you can have within the company. When you maximize your presence, you maximize your control of the company. With success comes influence and the ability to make a difference. A shrewd manager can successfully sway the direction of the company, how it treats the employees, and its profitability.

SUCCESS AT 30

The problem with success too soon is what to do next year or five years down the road. Over the years, I have seen a few very talented individuals quickly reach a pinnacle of success at a young age. As the manager, you need to monitor this type of worker's progress carefully. When a young employee has moved quickly to the top of the group, you can anticipate future problems. These problems can be annual salary, further advancement, and job satisfaction. All three of these issues will contribute to unhappy workers at some point in their careers.

Using the benchmark of 30 years of age, a person may have reached his or her peak earning potential for that job description. Understand that this person has been an exceptionally good employee, has been a person that you have counted on, and is uniquely qualified to be your number one candidate for advancement. The only problem is that this person lacks the years of experience necessary for advancement. After all, to be the best in a responsible position, a person must also have a wealth of practical experience. At 30 years of age, how seasoned can a young engineer, project manager, foreman, or technician be? This argument can be interpreted as discriminatory, but there is a lot to be said for wisdom from practical training. Ambitious young workers don't really want to hear this. They are anxious to keep climbing up the ladder of success. It is the manager's responsibility to regulate that progress, so that the employee will succeed.

This regulated growth is truly in the best interest of the employee. The logic behind having years of experience goes beyond immediate job advancement. You need to discuss the following points with the young employee:

- A 30-year old person still has more than 30 years to work.
- How much more can he or she possibly earn annually? And, will that salary be acceptable over the next 30 years?
- How did the person reach his or her peak earning potential at such a young age?
- Where does the person go from here? It's not that they don't deserve a salary increase, but you can make only so much at each job description/position.

- More importantly, how will the person accept the inherent limitations of any job position?

When success comes early, it creates adjustment problems for the successful individual. There may be no advancement openings for this person in the near future. As a result, the worker must choose to look elsewhere for advancement or accept the current position he or she holds within the company. The dilemma is a difficult but realistic fact of life, and it is the manager's job to introduce this reality to the employee's success. The issue is not to limit someone's professional growth; it is to bring an awareness to the employee that he or she is reaching a plateau that will require added time and commitment in order to climb to the next career level.

The same can be said for employees who are 40 years of age. At 40, a person has worked approximately 20 years, and most likely the person will work another 20 years. Therefore, you can consider the person to be at the midpoint of his or her professional life. Somewhere between 35 and 45, most workers have accomplished all of the advancements they will ever reach. They may or may not equate age with this career crest, because age is not the big issue; timing is the real issue. It is during this period of time when company enthusiasm, professional drive, and the excitement of the job begin to wane. People begin to realize that their job responsibilities are becoming more and more repetitive. They may be comfortable with their salary, although they would like to earn more. The juggling act to keep all the balls in the air may not be as important. Outside interests may now interfere with job advancement. Collectively, these issues are affecting the employees' work attitudes. As a manager, you should be cognizant of this type of person and should strive not to fall into this same pattern yourself. If you begin to lose your excitement for the job, workers, and clients, then your leadership ability becomes a company liability. A manager must always project a strong desire to lead. Whether you are a manager at 30, 40, or 50 years of age, you need to be the motivator.

How Far Can You Go?

To be a successful manager, you must understand the unspoken ground rules that go along with a promotion into management. For example, if the company is a "family-owned" firm, chances are you can get close to the top, but you will never be president of the company. Another example is if the company employees don't have the right to vote for the president of the company, and the present person in charge is relatively young. Your chances of "overthrowing" the president are usually very, very slim. With these kinds of ground rules, a manager must accept the things that can't be changed and concentrate on what can be changed. If you are going to succeed at your job, then you must accept the consensus and support those unwritten rules.

I have taken my own advice and focused on where am I going to be in one year and three years. Accept the responsibilities you are given, and learn from the on-the-job training you receive with practical experience. At the same time, I have recognized and appreciated the authority of those who are my superiors. I have never wished for or pursued ownership of my own company. As a result, I recognize that I must work within the corporate structure of someone else's company. Success can mean many things, and to me, job satisfaction is one of the most important forms of success. As a manager, if you are not excited about your responsibilities, the challenges that can occur, or even the mundane routines of the job, then success begins to diminish. Prior to that happening, you must follow the same advice you have given to other employees, and you must ask yourself the following questions:

- Where will I be in one year and three years?
- Will I still enjoy the job?

Don't wait until you are dissatisfied with your job and this displeasure is projected to the other workers.

For me, I have had a need to move a few times in my career. Only once did I leave because I was not happy with the company. It was the only time in my working career that I hated to get up in the morning and go to work. My first job change was strictly to work somewhere else. I worked at one firm for eight years, and I was anxious to see how other firms performed in the hvac/r industry. My last two employment moves were to enhance my knowledge of the entire hvac/r industry. The first of these two moves allowed me to expand from consulting engineering into designing, building, service, maintenance, and estimating. The next change expanded this sphere into construction management. With both previous employers, I also recognized that I had gone as far as I was going to go up the ladder of success. This awareness allowed me to be more focused on what I wanted to do with my professional life. With both of these past employments, I understood that if I was to advance any further within the company, then it would be into areas of responsibility for which I didn't want to be held accountable! You have to like your job, and if you don't, then you need to do something about the situation.

Managers are no different than those who advance from engineers-in-training to engineers, pipefitters to foremen, or technicians to lead engineers. Reaching a manager's position doesn't mean the end of your career. I was a manager of hvac/r engineering for four years and then chose to resign from the job. After four years, it wasn't fun anymore. The job became too people oriented and less concerned with design engineering. In addition, the next position up the ladder was all people related! My time was being consumed with employee reviews, hiring new workers to keep up with the expanded work load, employee benefits, and answering for the errors of others. I wanted to get back to designing my own jobs and making my own mistakes.

Managers must routinely assess their own job fulfillment. Don't get consumed with the success of the position and lose sight of your own personal goals. While you are busy helping the company and the employees, don't forget about yourself. Ask yourself where you are going to be in one year and three years. Effective managers should continually broaden their horizons, strengthen their education, and strive to know more about the issues closely related to the hvac/r industry. In the process, you need to have a plan to provide continued success. Know what you are good at, what you like to do, and how you are going to maintain your employment satisfaction.

Managers should recognize what they are good at and also what they think they could be good at. These skills usually complement one another and follow a natural progression from one to the other. For an example, if you are a design engineer, your next move may be to become an engineer. If you are an apprentice pipefitter, then your next advancement is to become a journeyman pipefitter. Successful managers need to think in these same terms. If you are a very good manager of engineering, then you may possibly be a candidate for manager of service. The methods for managing people are the same regardless. The hiring and firing process follows the same standards of performance. Annual reviews, managing by walking around, and time management are all integral to both manager positions. Another option may be to leave the firm and start your own company. An experienced manager will certainly have that option after he or she has fulfilled a number of years managing someone else's employees.

When you have come to grips with how far you can go as a manager, you must accept whatever limitations and/or potential you have and continue to be successful. Remember, if you plan to resign and/or move to a higher position, you will have fewer options and opportunities. Long-range planning is essential to job satisfaction. Don't let others know you are considering a job change. It can be demoralizing to others, even if it doesn't directly affect them. In addition, you don't want others to know, because daily problems might be blamed on your desire to leave. Having worked hard to be proficient at your job, you don't want others to take away from your success with accusations of poor performance. There is always some fellow employee who will be glad to see you leave. Usually, this person considers you to be in his or her way of advancing within the company. A good manager knows that there are always people who believe they can do a better job and are anxious to step forward. You just want to be out of their way when they step forward! Also remember that you can be replaced! The company has done well with you, and it can do well without you. You will know when it is time to leave. Try to make the transition with as little disruption as possible. If you have properly trained your predecessor, then he or she will continue the success as if you were never there.

The Manager-The Trainer

To succeed at helping others, you have to be able to help yourself. If you can't get ahead in the hvac/r industry, how do you expect to help others get ahead? The cornerstone to this success is time management. How you manage your time is integral to how those you are responsible for manage their time. Make the most of every workday, and project this attitude to others. Don't waste valuable hours. Learn the skills necessary for effective time management, and teach these same skills to others, so they too can benefit from maximized work performance. Continually plan your time well in advance, and use the many time management tools available to you, such as carrying your laptop computer with you, compiling lists, and listening to audiocassettes. Encourage others to plan accordingly, and make sure that they also listen to tapes!

Set the example by always using a "things to do" list, and expect the other workers to maintain a similar one page listing. Show them the benefits of prioritizing their tasks and keeping track of each item as it is completed. Also, stress to your employees the job satisfaction that is associated with the completion of each of these tasks. Help them realize the personal achievement of getting the job done. In addition, make sure everyone has developed a set of goals for the year. Help them with these company-related topics, and remind them of the need to continually monitor these milestones. All managers and employees should set a minimum of six annual goals, from which the company and they themselves will benefit.

If you are a first-time manager, be prepared to take charge. Plan your first, second, and third year strategy, and present it to your boss. Do your homework, and put everything down in writing. Don't leave anything to interpretation. Organize your thoughts, and practice your presentation long before you meet with your boss. After your plan is approved, educate the employees by sharing your plan with them. Show them there is a method to your madness! The more informed they are, the more help they will be to your three-year strategy. In addition, by sharing this blueprint for success, the employees inherently buy in to it! Show them how to be good businesspeople, and educate them on the realities of work, including the need to be financially responsible, to gain experience, to continue to learn, and to make maximum use of their time.

Maximize your leadership by setting the pace and setting the example. Selectively teach others to do the same. Together, you have increased your training staff, and this will help reinforce the educational process. As the manager, make the workplace a pleasant environment. People need to take pride in where they work. This can be in the main office, a branch office, a jobsite, or a service van. An organized and professional environment speaks for you, and your clients will see that the employees take pride in themselves, their work, and their company. Teach the other workers to appreciate this point of view. It can sometimes make the difference between your company and the competition.

The best managers also have a lot of tricks of the trade to share with the workers. Through experience and a devotion to detail, you can share this wealth of knowledge with others so the group will continue to excel. Keep good notes, and remember what worked and what didn't. Learn from your mistakes, but don't constantly rethink every decision that you make. Build upon your achievements, and share this information with the people for whom you are responsible. The more they know, the easier your job will be. Always remember to get the most out of your workers, for their benefit as well as the company's benefit. In the process, don't ever worry about losing your job to someone else. Focus on only those things that you can control. The more you improve your group through training, encouragement, and opportunity, the more valuable you will be to the firm. No one is going to replace a true leader. There are only a very few of these people to go around.

Learn to understand people. The better you understand people, the better chance you have of teaching and training them to be proficient at what they do. Recognize the signs that would indicate a negative attitude, such as routine complaining, poor work habits, slow to start working, and always blaming someone else. Inspire this type of worker to change, or encourage them to move on. Successful managers must continually strive to maximize their time and the time of others. Those who don't want to follow need to step aside. It is your job to keep the pessimists in the group to a minimum and maintain an upbeat attitude. Over the years, you will learn how to read people. This is an important skill to learn and share with your workers. Let them know their good points and those talents on which they need to work.

At the same time, observe other people in management. Learn what works for them and what doesn't. Sometimes people become absorbed in their success. This can develop into a double standard if they see themselves as more valuable than the rest of the workers. Everyone has a job to do and a responsibility to be the best at what their position description requires. I have had the luxury of working for both the best and worst managers. I have learned from each and have strived to do better. Managers must continually work to improve their own skills, so they can be a positive influence on a company. A boss who considers himself to be more valuable than the others is making a serious mistake. Usually, these leaders are the ones who are the least proficient at what they do. In addition, the employees can assess this marginal performance and form their own opinion of this manager's value to the company. The by-product is usually a company-wide bad attitude.

If you set corporate goals and maintain a series of milestones that document your progress, then you should be able to assess your value to the company. If you are successful with your own performance, you inherently demonstrate these accomplishments to the other employees. Without a word, the other workers can see that you are doing your part and that there is a common thread between their performance and the performance of the manager. In other words, they will see that there is no double standard!

WINDOWS OF OPPORTUNITY

For all of their time and effort, managers deserve a "one minute praising."[1] As a manager, you are responsible not only for yourself but for many others. This is more than a 40-hour-a-week job! Everything a manager does needs to be done in duplicate. What is good for you is good for the others. If your employees need to learn continually, then you need to learn continually. If you are going to set goals by which to measure your performance, then employees need to set goals by which to measure their performances. The incentives need to be a mutual venture, because you are all members of a team. If one person doesn't do his or her job, it reflects on the entire team. Being the team leader does not exclude you from participating. At the same time, not being the team leader does not excuse the workers from committing themselves 100%.

A moment of praising is something you are entitled to, even if it comes from yourself. Having continually looked out for the company and the employees, you should stop and reflect on those issues that have been a positive statement of your work. Managers owe it to themselves to step back and reflect on their achievements. Enjoy a few moments of quiet time when there are no phones, no people standing in your office doorway, no budget spreadsheets, etc. I consider this to be a sanity break from the pressures you put yourself through. It is necessary to take this quiet time to reflect on what you are doing, how well you are doing it, and where you are going in the process. Remember, while you are striving the create and maintain a financially successful company operation and individual job satisfaction for all the employees, you are entitled to these same benefits. There has to be something in it for you, something more than a financial incentive. Personal satisfaction is integral to good management.

Over the last 16 years, I have given up the position of manager twice. Both times were due, in part, to a loss of job satisfaction. If you don't possess the drive to continually lead others, then you need to step aside. A manager must work with and maintain a certain level of standards. If your focus begins to take you in another direction, then you need to follow that road and leave the one you are on. Sometimes people just realize that management is not the job for them, even after a number of years managing people. Similar to the way they advanced to a position as a manager, they recognize that it is time to move on to new opportunities. The trick to this move is to recognize the need early and plan your position change a year or two in advance.

Networking with your associates can help you get the word out discretely to others if there are no opportunities that interest you within the firm. By planning well in advance of your job change, a person can patiently wait for the right opening to come along. The higher position and salary you have, the longer it will take to find that next assignment. If you followed your own advice and pursued writing, publication of project case studies, awards, and membership in professional associations, then your goal should be easier to reach. Success reaps success. The more visible you have been in the past, the more opportunity you should have in the future.

An experienced manager has many windows of opportunity with which to work. The fact that you are a very good manager opens doors to other management roles in other departments. These other fields of the hvac/r business may be just what you need to sustain your drive for job satisfaction. A lateral change can bring new work incentive, because you are in a new arena of business. As you manage the employees, you can learn from them as well. A new work environment can be good for you for at least another four or five years. During that period of time, you can become proficient in the skills associated with that department. In addition, you attain a new trade to complement your most recent skills. The more you know about the hvac/r business, the more valuable you are to yourself and to others. Reflecting back on this knowledge in writing can be a nice diversion from your daily work. As you put down in writing your new experiences, you learn even more in the process.

A manager should continually strive to enhance his or her credentials by learning more about the industry. As you grow within the endless boundaries of the hvac/r industry, you can use this training to help with the training of others. Your goal should be to distinguish yourself as an authority, i.e., a person from whom others will want to learn. Because you continue to walk in their shoes, you can offer a tremendous learning benefit to those you manage.

SUMMARY

In order to be a very good manager, you must be a very good student. The successful manager constantly practices what he or she preaches. Never sit back and believe you have finally made it! Every day a manager has to take charge, provide the work ethic, excitement, and the business savvy to succeed. Once you let up on the drive to be the best, then someone else will come along and take that title. Over the years, I have seen various companies have their "day in the sun." In time, another company would come along and work just a little harder and be a little more persistent in order to cast a shadow over the other firm. It is up to the manager to be the inspiration to the workers, so that your firm can stay out in front the longest.

Success can be a double-edged sword that offers you the chance to learn from both good performances and bad performances. Don't ever miss an opportunity to question "why" or "why not" when a project is a success or a failure; when a person hasn't followed procedure; or when a worker isn't satisfied with his or her job. Continually observe what is happening around you, so you can continue to improve the company and its employees. Never waste time feeling bad for something you did or didn't do. I have never met a person in this business yet who did something wrong on purpose. Errors happen, and it is your job to correct them, not judge them. The hvac/r industry is made up of credible, hard working individuals and a number of equally credible professional associations. You need to focus on the positive.

At the same time, you should reflect back on your own success within the business. Taking the time to evaluate your performance and appreciate these achievements can reinforce your commitment to teach others. A manager routinely needs to reflect back on his or her own background when helping others in the group. You can be a wealth of information to less experienced workers. While pondering what they should be doing, they can appreciate what you have already done before them. This can be quite helpful to the younger employees who are in a trainee or apprentice position. A veteran manager can have a significant influence over a new employee. Hopefully it will be an influence that will help the person become an excellent worker.

Finally, when managers tire of the day-to-day tasks associated with being in charge, they should plan to redirect their own energy towards newer challenges. After all, they got to where they are in management through a continuing process of advancement. Now may be the time to step aside and let someone else lead while you focus on other endeavors. If you have to work until you are 62 or 65, you might as well enjoy what you are doing in the process!

NOTES

[1]   Blanchard, Kenneth, Ph.D. and Spencer Johnson, M.D., *The One Minute Manager*, William Morrow and Co., Inc., New York, 1982.

Appendix
Sample Concept Package

For engineering, one of the most effective time management tools is the ability to conceptualize systems. To enhance this unique skill, a design engineer should learn to organize his or her thoughts using a 17 section specification checklist laid out on 8-1/2" x 11" or 11" x 17" sketch pads. Why document system drawings and narratives on sheets any larger when sheets of these sizes can efficiently provide the customer with the pertinent data?

Conceptualizing occurs within the first 5% of any project, and it is during this initial phase that the hvac/r engineer is called upon to develop a brief, concise overview of the entire scope of work. The Sample Concept Package follows the format of the Construction Specification Institute, Division 01000 through 17000. This standardized process is convenient when documenting the design data, as well as reviewing the documentation. By maintaining this information in booklet form, the engineer can quickly check off the prescriptive data base. This exercise can be completed at the project site and/or completed while returning in-flight, on a train, bus, or even in a coffee shop prior to returning to the office.

At the completion of this exercise, the Sample Concept Package will include one-line flow diagrams, preliminary equipment selections, and the 17 section specification. In addition, these flow diagrams should be copied to incorporate electrical data, automatic controls, operating management strategy, equipment weights, and access and servicing clearances. The design engineer should also make copies of the catalog equipment cut-sheets. By using these "tricks of the trade," an experienced engineer can conceptualize an hvac/r design in a time-managed manner.

Sample Concept Package		
Project Title	**Project Scope**	
Division 01000 - General Requirements	**A**	**B**
Allowances		
1. Architect and Engineer Fees		
2. Construction Management Fees		
3. Contingencies		
4. Performance Bond		
Equipment and Facilities		
1. Rental, etc.		
2. Field Office		
3. Dumpster		
4.		
Shop Drawing Submittals		
1. All Trades		
Operation and Maintenance Manuals		
1. Equipment Manuals		
2. Computerized Maintenance Manuals		
3. O and M Training Videos		
Coordination Drawings		
1. CAD Compatible		
Equipment and System Identification		
1. Section #11000 Equipment		
2. Section #14000 Equipment		
3. Sections #15400 through #17000		

Project Title	Project Scope	
Division 01000 — General Requirements (Cont.)	A	B
Testing Services		
1. Pressure Test Pipe Systems		
2.		
3.		
Cleaning		
1.		
2.		
Training		
1. Job Specific Training		
2. Manufacturer's Training		
Start-up		
1. Subcontractor Sign-off		
2. Submission of Testing Reports		
3. Architect/Engineer Sign-off		
4. Acoustic Reading Report		
5. Equipment Performance Test		
6. System Performance Test		
Close-out		
1. Warranty		
2. Inventory		
3.		
4.		
Division 15500 — Hvac/r		
General Requirements		
1. Shop Drawing Submittals		
2. O and M Manuals		
3. Coordination Drawing		
4. Start-up		
5. Close-out		

Project Title	Project Scope	
Division 15500 — Hvac/r (Cont.)	A	B
Systems		
1. Hvac/r for:		
2. Hvac/r for:		
3. Central Heating Plant		
4. Central Cooling Plant		
5. General Exhaust		
6. Fume Hood Exhaust		
7. Smoke Exhaust		
8.		
9.		
10.		
11.		
12.		
Testing		
1. Pressure Test Sheet Metal Systems		
2. Cleaning — Air Systems		
3. Air and Water Balancing		
4. Control Calibration		
Identification		
1. Equipment		
2. Pipe		
3. Sheet Metal		
4. Color Coding		
5. Valve Tags		
Equipment		
1. Boilers		
2. Chillers		
3. Pumps		
4. Air Handling Units		
5. Fans		
6. Coil		

Project Title	Project Scope	
Division 15500 — Hvac/r (Cont.)	**A**	**B**
7. Filters		
8. Heat Exchangers		
9. Air Terminals		
10. Water Terminals		
11. Humidifiers		
12. Insulation		
13.		
14.		
Automatic Temperature Controls		
1.		
2.		